SOLID STATE CHEMISTRY: AN INTRODUCTION

Edited by
Jean Annerie Hernandez

Delve Publishing

Solid State Chemistry: An Introduction

Edited by Jean Annerie Hernandez

Edition 2018

Delve Publishing

Published by

Delve Publishing LLC

708 3rd Avenue, 6th Floor

New York

NY 10801

United States of America

www.delvepublishing.com

ISBN: 978-1-68095-731-0 (Hardback)

Printed in United States of America

Library of Congress Control Number: 2016920411

About the Editor

Jean Annerie Hernandez

Jean Annerie took up a Bachelor's degree in Chemistry in Central Luzon State University before she decided to fulfill her lifelong dream of becoming a creative writer and editor. She has ghostwritten and copyedited fictional and non-fictional books for three years, and is currently working on various client-based projects in Singapore, China and USA

Table of Contents

5 MAGNETISM

6 SUPERCONDUCTIVITY

7 ALLOY: STRUCTURE AND PROPERTIES

8 METALLURGY

List of Abbreviations

ABS	acrylonitrile butadiene styrene
APS	Air plasma spraying
ARPES	angle resolved photoemission spectroscopy
CAPS	Controlled atmosphere plasma spraying
GSP	Gas-stabilized plasma
HVAF	High velocity air fuel
HVOF	High velocity oxygen fuel spraying
MRI	magnetic resonance imaging
MSNs	Mesoporous silica nanoparticles
NMR	nuclear magnetic resonance
PVDF	polymer polyvinylidene fluoride
PEM	proton exchange membrane
RCC	Reinforced Carbon-Carbon
SAXS	small-angle X-ray scattering
SOFC	solid oxide fuel cells
SSC	Superconducting Super Collider
IUPAC	Union of Pure and Applied Chemistry
VPS	Vacuum plasma spraying
WSP	Water-stabilized plasma

Preface

Solid is one of the three main states of matter, along with liquid and gas. Matter is the "stuff" of the universe, the atoms, molecules and ions that make up all physical substances. In a solid, these particles are packed closely together and are not free to move about within the substance. Molecular motion for the particles in a solid is confined to very small vibrations of the atoms around their fixed positions; therefore, solids have a fixed shape that is difficult to change. Solids also have a definite volume; that is, they keep their size no matter how you try to change them. Materials chemistry produces an understanding of material properties, in particular by creating chemical correlations. This is based on four steps ranging from the design of a material, the synthesis and characterization up to potential applications. While the latter is more about the engineering disciplines, the first three steps are based mainly on chemical and physical principles. One focus of the materials chemistry therefore constitutes the methodical combination of chemical synthesis and characterization together with topics from materials science. Because of its direct relevance to products of commerce, solid state inorganic chemistry has been strongly driven by technology. Progress in the field has often been fueled by the demands of industry, well ahead of purely academic curiosity. Applications discovered in the 20th century include zeolite and platinum-based catalysts for petroleum processing in the 1950s, high-purity silicon as a core component of microelectronic devices in the 1960s, and "high temperature" superconductivity in the 1980s.

Content Coverage

Chapter One presents the introduction of crystal structure. The crystalline state of matter is the state with the highest order, i.e., with very high internal correlations and at the greatest distance range.

Chapter Two focuses on the techniques used to determine the arrangement of atoms of a crystalline solid in three dimensional space.

Chapter Three outlines the structure and properties of solids. It is characterized by structural rigidity and resistance to changes of shape or volume.

Chapter Four highlights about the ionic conductivity. In solids, ions typically occupy fixed positions in the crystal lattice and do not move. However, ionic conduction can occur, especially as the temperature increases.

Chapter Five focuses on the phenomena of magnetism that are mediated by magnetic fields. Electric currents and the magnetic moments of elementary particles give rise to a magnetic field, which acts on other currents and magnetic moments.

Chapter Six emphasizes on the phenomenon of superconductivity of exactly zero electrical resistance and expulsion of magnetic flux fields occurring in certain materials, called superconductors, when cooled below a characteristic critical temperature.

Chapter Seven focuses on alloy structure and properties. The alloy constituents are usually measured by mass. Alloys are usually classified as substitutional or interstitial alloys, depending on the atomic arrangement that forms the alloy.

Chapter Eight presents the study of the physical and chemical behavior of metallic elements, their intermetallic compounds, and their mixtures, which are called alloys. Metallurgy is also the technology of metals: the way in which science is applied to the production of metals, and the engineering of metal components for usage in products for consumers and manufacturers.

CHAPTER 1

Crystal Structures

Chapter Contents

- **Introduction**
- **Crystal Structure (Microscopic)**
- **Crystal Faces and Shapes**
- **Occurrence in Nature**
- **Crystallization**
- **Defects, Impurities, and Twinning**
- **Chemical Bonds**
- **Quasicrystals**
- **Crystallography**
- **Crystal Growth**

A crystal structure is composed of a unit cell, a set of atoms arranged in a particular way; which is periodically repeated in three dimensions on a lattice.

The crystalline state of matter is the state with the highest order, ie, with very high internal correlations and at the greatest distance range. This is reflected in their properties: anisotropic and discontinuous. Crystals usually appear as unadulterated, homogenous and with well-defined geometric shapes (habits) when they are well-formed. However, as we say in Spanish, "the habit does not make the monk" (clothes do not make the man) and their external morphology is not sufficient to evaluate the crystallinity of a material.

INTRODUCTION

A **crystal** or **crystalline solid** is a solid material whose constituents (such as atoms, molecules, or ions) are arranged in a highly ordered microscopic structure, forming a crystal lattice that extends in all directions. In addition, macroscopic single crystals are usually identifiable by their geometrical shape, consisting of flat faces with specific, characteristic orientations. The scientific study of crystals and crystal formation is known as crystallography. The process of crystal formation via mechanisms of crystal growth is called crystallization or solidification.

A crystal of amethyst quartz.

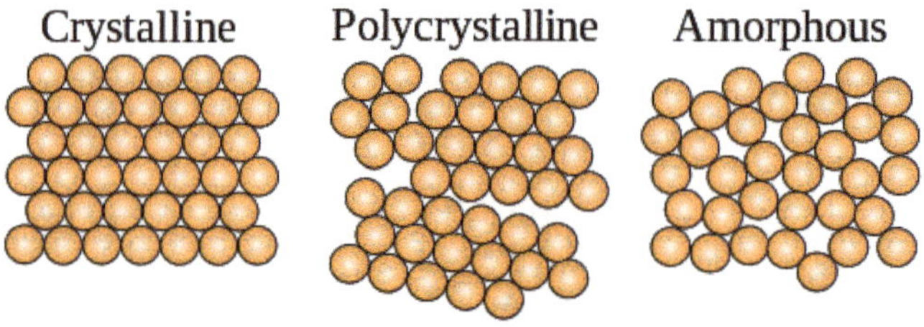

Microscopically, a single crystal has atoms in a near-perfect periodic arrangement; a polycrystal is composed of many microscopic crystals (called "crystallites orgrains); and an amorphous solid (such as glass) has no periodic arrangement even microscopically.

The word *crystal* derives from the Ancient Greek word κρύσταλλος (*krustallos*), meaning both "ice" and "rock crystal", from κρύος (*kruos*), "icy cold, frost".

Examples of large crystals include snowflakes, diamonds, and table salt. Most inorganic solids are not crystals but polycrystals, i.e. many microscopic crystals fused together into a single solid. Examples of polycrystals include most metals, rocks, ceramics, and ice. A third category of solids is amorphous solids, where the atoms have no periodic structure whatsoever. Examples of amorphous solids include glass, wax, and many plastics.

Crystals are often used in pseudoscientific practices such as crystal therapy, and, along with gemstones, are sometimes associated with spellwork in Wiccan beliefs and related religious movements.

CRYSTAL STRUCTURE (MICROSCOPIC)

Halite (table salt, NaCl): Microscopic and Macroscopic

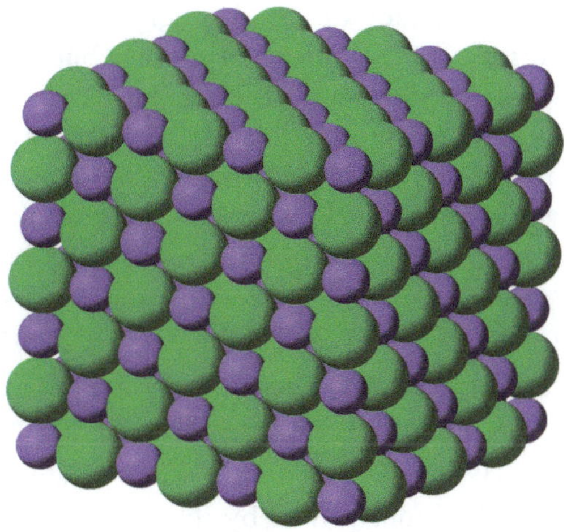

Microscopic structure of a halite crystal. (Purple is sodium ion, green is chlorine ion.) There is cubic symmetry in the atoms' arrangement.

Macroscopic (~16cm) halite crystal. The right-angles between crystal faces are due to the cubic symmetry of the atoms' arrangement.

The scientific definition of a "crystal" is based on the microscopic arrangement of atoms inside it, called the crystal structure. A crystal is a solid where the atoms form a periodic arrangement. (Quasicrystals are an exception, see below.)

Not all solids are crystals. For example, when liquid water starts freezing, the phase change begins with small ice crystals that grow until they fuse, forming a *polycrystalline* structure. In the final block of ice, each of the small crystals (called "crystallites orgrains) is a true crystal with a periodic arrangement of atoms, but the whole polycrystal does *not* have a periodic arrangement of atoms, because the periodic pattern is broken at the grain boundaries. Most macroscopic inorganic solids are polycrystalline, including almost all metals, ceramics, ice, rocks, etc. Solids that are neither crystalline nor polycrystalline, such as glass, are called *amorphous solids*, also called glassy, vitreous, or noncrystalline. These have no periodic order, even microscopically. There are distinct differences between crystalline solids and amorphous solids: most notably, the process of forming a glass does not release the latent heat of fusion, but forming a crystal does.

A crystal structure (an arrangement of atoms in a crystal) is characterized by its *unit cell*, a small imaginary box containing one or more atoms in a specific spatial arrangement. The unit cells are stacked in three-dimensional space to form the crystal.

The symmetry of a crystal is constrained by the requirement that the unit cells stack perfectly with no gaps. There are 219 possible crystal symmetries, called crystallographic space groups. These are grouped into 7 crystal systems, such as cubic crystal system (where the crystals may form cubes or rectangular boxes, such as halite shown at right) or hexagonal crystal system (where the crystals may form hexagons, such as ordinary water ice).

CRYSTAL FACES AND SHAPES

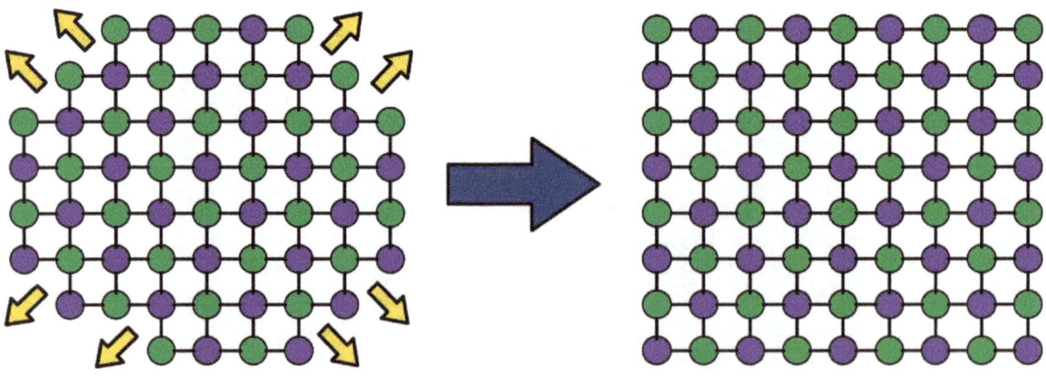

As a halite crystal is growing, new atoms can very easily attach to the parts of the surface with rough atomic-scale structure and many dangling bonds. Therefore, these parts of the crystal grow out very quickly (yellow arrows). Eventually, the whole surface consists of smooth, stable faces, where new atoms cannot as easily attach themselves.

Crystals are commonly recognized by their shape, consisting of flat faces with sharp angles. These shape characteristics are not *necessary* for a crystal—a crystal is scientifically defined by its microscopic atomic arrangement, not its macroscopic shape—but the characteristic macroscopic shape is often present and easy to see.

Euhedral crystals are those with obvious, well-formed flat faces. Anhedral crystals do not, usually because the crystal is one grain in a polycrystalline solid.

The flat faces (also called facets) of a euhedral crystal are oriented in a specific way relative to the underlying atomic arrangement of the crystal: they are planes of relatively low Miller index. This occurs because some surface orientations are more stable than others (lower surface energy). As a crystal grows, new atoms attach easily to the rougher and less stable parts of the surface, but less easily to the flat, stable surfaces. Therefore, the flat surfaces tend to grow larger and smoother, until the whole crystal surface consists of these plane surfaces. (See diagram on right.)

One of the oldest techniques in the science of crystallography consists of measuring the three-dimensional orientations of the faces of a crystal, and using them to infer the underlying crystal symmetry.

A crystal›s habit is its visible external shape. This is determined by the crystal structure (which restricts the possible facet orientations), the specific crystal chemistry and bonding (which may favor some facet types over others), and the conditions under which the crystal formed.

OCCURRENCE IN NATURE

Ice crystals.

Fossil shell with calcite crystals.

Rocks

By volume and weight, the largest concentrations of crystals in the Earth are part of its solid bedrock. Crystals found in rocks typically range in size from a fraction of a millimetre to several centimetres across, although exceptionally large crystals are occasionally found. As of 1999, the world's largest known naturally occurring crystal is a crystal of beryl from Malakialina, Madagascar, 18 m (59 ft) long and 3.5 m (11 ft) in diameter, and weighing 380,000 kg (840,000 lb).

Some crystals have formed by magmatic and metamorphic processes, giving origin to large masses of crystalline rock. The vast majority of igneous rocks are formed from molten magma and the degree of crystallization depends primarily on the conditions under which they solidified. Such rocks as granite, which have cooled very slowly and under great pressures, have completely crystallized; but many kinds of lava were poured out at the surface and cooled very rapidly, and in this latter group a small amount of amorphous or glassy matter is common. Other crystalline rocks, the metamorphic rocks such as marbles, mica-schists and quartzites, are recrystallized. This means that they were at first fragmental rocks like limestone, shale and sandstone and have never been in a molten condition nor entirely in solution, but the high temperature and pressure conditions of metamorphism have acted on them by erasing their original structures and inducing recrystallization in the solid state.

Other rock crystals have formed out of precipitation from fluids, commonly water, to form druses or quartz veins. The evaporites such as halite, gypsum and some limestones have been deposited from aqueous solution, mostly owing to evaporation in arid climates.

Ice

Water-based ice in the form of snow, sea ice and glaciers is a very common manifestation of crystalline or polycrystalline matter on Earth. A single snowflake is typically a single crystal, while an ice cube is a polycrystal.

Organigenic Crystals

Many living organisms are able to produce crystals, for example calcite and aragonite in the case of most molluscs or hydroxylapatite in the case of vertebrates.

Polymorphism and Allotropy

The same group of atoms can often solidify in many different ways. Polymorphism is the ability of a solid to exist in more than one crystal form. For example, water ice is ordinarily found in the hexagonal form Ice I_h, but can also exist as the cubic Ice I_c, the rhombohedral ice II, and many other forms. The different polymorphs are usually called different *phases*.

In addition, the same atoms may be able to form noncrystalline phases. For example, water can also form amorphous ice, while SiO_2 can form both fused silica (an amorphous glass) and quartz (a crystal). Likewise, if a substance can form crystals, it can also form polycrystals.

For pure chemical elements, polymorphism is known as allotropy. For example, diamond and graphite are two crystalline forms of carbon, while amorphous carbon is a noncrystalline form. Polymorphs, despite having the same atoms, may have wildly different properties. For example, diamond is among the hardest substances known, while graphite is so soft that it is used as a lubricant.

Polyamorphism is a similar phenomenon where the same atoms can exist in more than one amorphous solid form.

CRYSTALLIZATION

Vertical cooling crystallizer in a beet sugar factory.

Crystallization is the process of forming a crystalline structure from a fluid or from materials dissolved in a fluid. (More rarely, crystals may be deposited directly from gas; see thin-film deposition and epitaxy.)

Crystallization is a complex and extensively-studied field, because depending on the conditions, a single fluid can solidify into many different possible forms. It can form a single crystal, perhaps with various possible phases, stoichiometries, impurities, defects, and habits. Or, it can form a polycrystal, with various possibilities for the size, arrangement, orientation, and phase of its grains. The final form of the solid is determined by the conditions under which the fluid is being solidified, such as the chemistry of the fluid, the ambient pressure, the temperature, and the speed with which all these parameters are changing.

Specific industrial techniques to produce large single crystals (called *boules*) include the Czochralski process and the Bridgman technique. Other less exotic methods of crystallization may be used, depending on the physical properties of the substance, including hydrothermal synthesis, sublimation, or simply solvent-based crystallization.

Large single crystals can be created by geological processes. For example, selenite crystals in excess of 10 meters are found in the Cave of the Crystals in Naica, Mexico. For more details on geological crystal formation, see above.

Crystals can also be formed by biological processes, see above. Conversely, some organisms have special techniques to *prevent* crystallization from occurring, such as antifreeze proteins.

DEFECTS, IMPURITIES, AND TWINNING

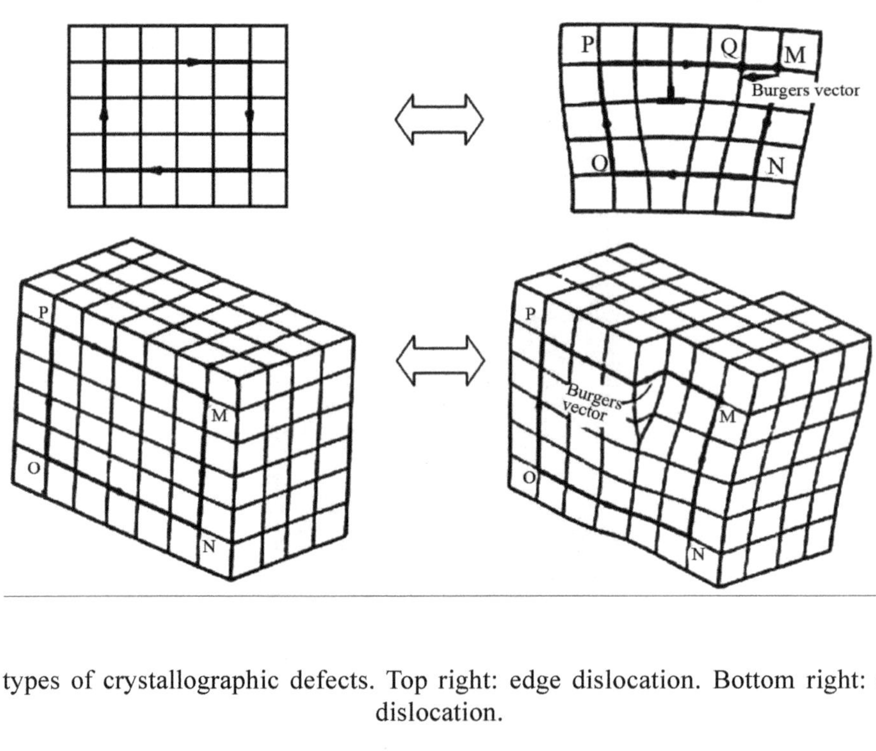

Two types of crystallographic defects. Top right: edge dislocation. Bottom right: screw dislocation.

An *ideal* crystal has every atom in a perfect, exactly repeating pattern. However, in reality, most crystalline materials have a variety of crystallographic defects, places where the crystal›s pattern is interrupted. The types and structures of these defects may have a profound effect on the properties of the materials.

A few examples of crystallographic defects include vacancy defects (an empty space where an atom should fit), interstitial defects (an extra atom squeezed in where it does not fit), and dislocations (see figure at right). Dislocations are especially important in materials science, because they help determine the mechanical strength of materials.

Another common type of crystallographic defect is an impurity, meaning that thewrong type of atom is present in a crystal. For example, a perfect crystal of diamond would only contain carbon atoms, but a real crystal might perhaps contain a few boron atoms as well. These boron impurities change the diamond›s color to slightly blue. Likewise, the only difference between ruby and sapphire is the type of impurities present in a corundum crystal.

Twinned pyrite crystal group.

In semiconductors, a special type of impurity, called a dopant, drastically changes the crystal›s electrical properties. Semiconductor devices, such as transistors, are made possible largely by putting different semiconductor dopants into different places, in specific patterns.

Twinning is a phenomenon somewhere between a crystallographic defect and a grain boundary. Like a grain boundary, a twin boundary has different crystal orientations on its two sides. But unlike a grain boundary, the orientations are not random, but related in a specific, mirror-image way.

Mosaicity is a spread of crystal plane orientations. A mosaic crystal is supposed to consist of smaller crystalline units that are somewhat misaligned with respect to each other.

CHEMICAL BONDS

In general, solids can be held together by various types of chemical bonds, such as metallic bonds, ionic bonds, covalent bonds, van der Waals bonds, and others. None of these are necessarily crystalline or non-crystalline. However, there are some general trends as follows.

Metals are almost always polycrystalline, though there are exceptions like amorphous metal and single-crystal metals. The latter are grown synthetically. (A microscopically-small piece of metal may naturally form into a single crystal, but larger pieces generally do not.) Ionic compound materials are usually crystalline or polycrystalline. In practice, large salt crystals can be created by solidification of a molten fluid, or by crystallization out of a solution. Covalently bonded solids (sometimes called covalent network solids) are also very common, notable examples being diamond and quartz. Weak van der Waals forces also help hold together certain crystals, such as crystalline molecular solids, as well as the interlayer bonding in graphite. Polymer materials generally will form crystalline regions, but the lengths of the molecules usually prevent complete crystallization—and sometimes polymers are completely amorphous.

QUASICRYSTALS

The material holmium–magnesium–zinc (Ho–Mg–Zn) forms quasicrystals, which can take on the macroscopic shape of a dodecahedron. (Only a quasicrystal, not a normal crystal, can take this shape.) The edges are 2 mm long.

A quasicrystal consists of arrays of atoms that are ordered but not strictly periodic. They have many attributes in common with ordinary crystals, such as displaying a discrete pattern in x-ray diffraction, and the ability to form shapes with smooth, flat faces.

Quasicrystals are most famous for their ability to show five-fold symmetry, which is impossible for an ordinary periodic crystal (see crystallographic restriction theorem).

The International Union of Crystallography has redefined the term "crystal" to include both ordinary periodic crystals and quasicrystals ("any solid having an essentially discrete diffraction diagram").

Quasicrystals, first discovered in 1982, are quite rare in practice. Only about 100 solids are known to form quasicrystals, compared to about 400,000 periodic crystals known in 2004. The 2011 Nobel Prize in Chemistry was awarded to Dan Shechtman for the discovery of quasicrystals.

Special properties from anisotropy

Crystals can have certain special electrical, optical, and mechanical properties that glass and polycrystals normally cannot. These properties are related to the anisotropy of the crystal, i.e. the lack of rotational symmetry in its atomic arrangement. One such property is the piezoelectric effect, where a voltage across the crystal can shrink or stretch it. Another is birefringence, where a double image appears when looking through a crystal. Moreover, various properties of a crystal, including electrical conductivity, electrical permittivity, and Young›s modulus, may be different in different directions in a crystal. For example, graphite crystals consist of a stack of sheets, and although each individual sheet is mechanically very strong, the sheets are rather loosely bound to each other. Therefore, the mechanical strength of the material is quite different depending on the direction of stress.

Not all crystals have all of these properties. Conversely, these properties are not quite exclusive to crystals. They can appear in glasses or polycrystals that have been made anisotropic by working or stress—for example, stress-induced birefringence.

CRYSTALLOGRAPHY

Crystallography is the science of measuring the crystal structure (in other words, the atomic arrangement) of a crystal. One widely used crystallography technique is X-ray diffraction. Large numbers of known crystal structures are stored in crystallographic databases.

Gallery

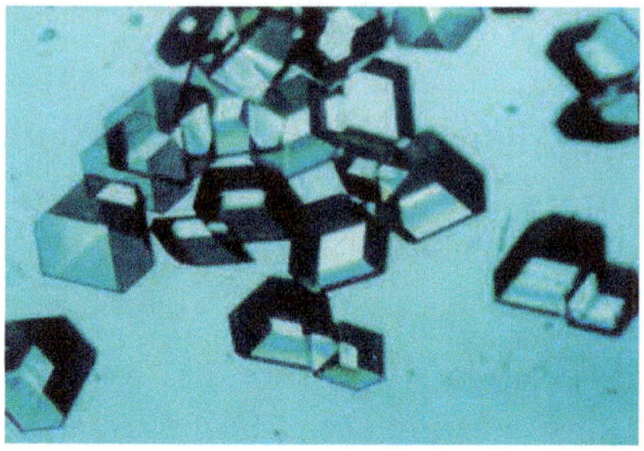

Insulin crystals grown in earth orbit.

Hoar frost: A type of ice crystal (picture taken from a distance of about 5 cm).

Gallium, a metal that easily forms large crystals.

An apatite crystal sits front and center on cherry-red rhodochroite rhombs, purple fluorite cubes, quartz and a dusting of brass-yellow pyrite cubes.

Boules of silicon, like this one, are an important type of industrially-produced single crystal.

A specimen consisting of a bornite-coated chalcopyrite crystal nestled in a bed of clear quartz crystals and lustrous pyrite crystals. The bornite-coated crystal is up to 1.5 cm across.

CRYSTAL GROWTH

Quartz is one of the several thermodynamically stable crystalline forms of silica, SiO_2.

A crystal is a solid material whose constituent atoms, molecules, or ions are arranged in an orderly repeating pattern extending in all three spatial dimensions. **Crystal growth** is a major stage of a crystallization process, and consists in the addition of new atoms, ions, or polymer strings into the characteristic arrangement of a crystalline Bravais lattice. The growth typically follows an initial stage of either homogeneous or heterogeneous (surface catalyzed) nucleation, unless aseed crystal, purposely added to start the growth, was already present.

The action of crystal growth yields a crystalline solid whose atoms or molecules are typically close packed, with fixed positions in space relative to each other. The crystalline state of matter is characterized by a distinct structural rigidity and virtual resistance to deformation (i.e. changes of shape and/or volume). Most crystalline solids have high values both of Young›s modulus and of the shear modulus of elasticity. This contrasts with most liquids or fluids, which have a low shear modulus, and typically exhibit the capacity for macroscopic viscous flow.

Crystalline solids are typically formed by cooling and solidification from the liquid state. According to the Ehrenfest classification of first-order phase transitions, there is a discontinuous change in volume (and thus a discontinuity in the slope or first derivative with respect to temperature, dV/dT) at the melting point. Within this context, the crystal and melt are distinct phases with an interfacial discontinuity having a surface of tension with a positive surface energy. Thus, a metastable parent phase is always stable with respect to the nucleation of small embryos or droplets from a daughter phase, provided it has a positive surface of tension. Such first-order transitions must proceed by the advancement of an interfacial region whose structure and properties vary discontinuously from the parent phase.

The process of nucleation and growth generally occurs in two different stages. In the first nucleation stage, a small nucleus containing the newly forming crystal is created. Nucleation occurs relatively slowly as the initial crystal components must impinge on each other in the correct orientation and placement for them to adhere and form the crystal. After crystal nucleation, the second stage of growth rapidly ensues. Crystal growth spreads outwards from the nucleating site. In this faster process, the elements which form the motif add to the growing crystal in a prearranged system, the crystal lattice, started in crystal nucleation. As first pointed out by Frank, perfect crystals would only grow exceedingly slowly. Real crystals grow comparatively rapidly because they contain dislocations (and other defects), which provide the necessary growth points, thus providing the necessary catalyst for structural transformation and long-range order formation.

Discontinuity

The conditions of a homogeneous environment are often approximated to but rarely ever realized. Crystal growth always involves some form of transport of matter or heat (or both). And homogeneous conditions for the transport process can only exist for spherical, cylindrical, or infinite plane surfaces. A polyhedral crystal cannot grow (remaining polyhedral) with uniform levels of supersaturation (or supercooling) over its faces. In general, the supersaturation is greatest at its corners. This refutes the assumption that the growth rate is a function of orientation and local supersaturation.

Thus, the crystal face must grow as a whole. The growth rate of the entire face is determined by the driving force (level of supersaturation) at the point of emergence of the predominant point of growth (e.g. a dislocation, a foreign particle acting as catalyst, or crystal twin). The defect-free habit face can thus resist a finite level of supersaturation without any growth at all.

Josiah Willard Gibbs was the first to point out that in the growth of a perfect crystal, the first derivative of the free energy with respect to mass becomes periodically undefinable — at each time that an additional layer on the crystal face is completed. There is discontinuity in the chemical potential at each such point.

In one sense, the crystal can then be in equilibrium with environments having a range of chemical potentials. In another sense, it is not in equilibrium. There are available states of lower free energy. But any free energy barrier must be passed by a fluctuation, or nucleation process, in order to access it. The fundamental thermodynamic effect of a screw dislocation is to eliminate this discontinuity in the chemical potential, by making it impossible to ever complete a single crystal face.

Nucleation

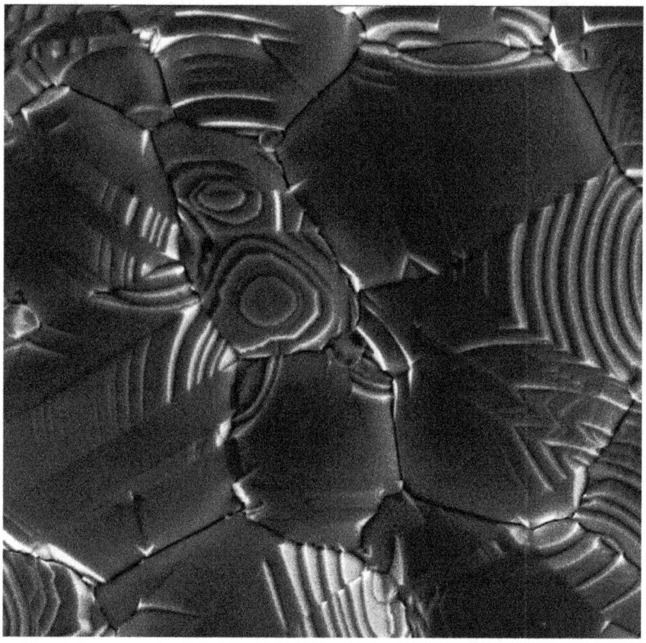

Silver crystal growing on a ceramic substrate.

Nucleation can be either homogeneous, without the influence of foreign particles, or heterogeneous, with the influence of foreign particles. Generally, heterogeneous nucleation takes place more quickly since the foreign particles act as a scaffold for the crystal to grow on, thus eliminating the necessity of creating a new surface and the incipient surface energy requirements.

Heterogeneous nucleation can take place by several methods. Some of the most typical are small inclusions, or cuts, in the container the crystal is being grown on. This includes scratches on the sides and bottom of glassware. A common practice in crystal growing is to add a foreign substance, such as a string or a rock, to the solution, thereby providing nucleation sites for facilitating crystal growth and reducing the time to fully crystallize.

The number of nucleating sites can also be controlled in this manner. If a brand-new piece of glassware or a plastic container is used, crystals may not form because the container surface is too smooth to allow heterogeneous nucleation. On the other hand, a badly scratched container will result in many lines of small crystals. To achieve a moderate number of medium-sized crystals, a container which has a few scratches works best. Likewise, adding small previously made crystals, or seed crystals, to a crystal growing project will provide nucleating sites to the solution. The addition of only one seed crystal should result in a larger single crystal.

Some important features during growth are the arrangement, the origin of growth, the interface form (important for the driving force), and the final size. When origin of growth is only in one direction for all the crystals, it can result in the material becoming very anisotropic (different properties in different directions). The interface form determines the additional free energy for each volume of crystal growth.

Lattice arrangement in metals often takes the structure of body centered cubic, face centered cubic, or hexagonal close packed. The final size of the crystal is important for mechanical properties of materials. (For example, in metals it is widely acknowledged that large crystals can stretch further due to the longer deformation path and thus lower internal stresses.).

Mechanisms of Growth

An example of the cubic crystals typical of the rock-salt structure.

The interface between a crystal and its vapor can be molecularly sharp at temperatures well below the melting point. An ideal crystalline surface grows by the spreading of single layers, or equivalently, by the lateral advance of the growth steps bounding the layers. For perceptible growth rates, this mechanism requires a finite driving force (or degree of supercooling) in order to lower the nucleation barrier sufficiently for nucleation to occur by means of thermal fluctuations. In the theory of crystal growth from the melt, Burton and Cabrera have distinguished between two major mechanisms:

- **Non-uniform lateral growth**. The surface advances by the lateral motion of steps which are one interplanar spacing in height (or some integral multiple thereof). An element of surface undergoes no change and does not advance normal to itself except during the passage of a step, and then it advances by the step height. It is useful to consider the step as the transition between two adjacent regions of a surface which are parallel to each other and thus identical in configuration — displaced from each other by an integral number of lattice planes. Note here the distinct possibility of a step in a diffuse surface, even though the step height would be much smaller than the thickness of the diffuse surface.

- **Uniform normal growth**. The surface advances normal to itself without the necessity of a stepwise growth mechanism. This means that in the presence of a sufficient thermodynamic driving force, every element of surface is capable of a continuous change contributing to the advancement of the interface. For a sharp or discontinuous surface, this continuous change may be more or less uniform over large areas each successive new layer. For a more diffuse surface, a continuous growth mechanism may require change over several successive layers simultaneously.

Non-uniform lateral growth is a geometrical motion of steps — as opposed to motion of the entire surface normal to itself. Alternatively, uniform normal growth is based on the time sequence of an element of surface. In this mode, there is no motion or change except when a step passes via a continual change. The prediction of which mechanism will be operative under any set of given conditions is fundamental to the understanding of crystal growth. Two criteria have been used to make this prediction:

- Whether or not the surface is **diffuse**. A diffuse surface is one in which the change from one phase to another is continuous, occurring over several atomic planes. This is in contrast to a sharp surface for which the major change in property (e.g. density or composition) is discontinuous, and is generally confined to a depth of one interplanar distance.
- Whether or not the surface is **singular**. A singular surface is one in which the surface tension as a function of orientation has a pointed minimum. Growth of singular surfaces is known to requires steps, whereas it is generally held that non-singular surfaces can continuously advance normal to themselves.

Driving Force

Consider next the necessary requirements for the appearance of lateral growth. It is evident that the lateral growth mechanism will be found when any area in the surface can reach a metastable equilibrium in the presence of a driving force. It will then tend to remain in such an equilibrium configuration until the passage of a step. Afterward, the configuration will be identical except that each part of the step but will have advanced by the step height. If the surface cannot reach equilibrium in the presence of a driving force, then it will continue to advance without waiting for the lateral motion of steps.

Thus, Cahn concluded that the distinguishing feature is the ability of the surface to reach an equilibrium state in the presence of the driving force. He also concluded that for every surface or interface in a crystalline medium, there exists a critical driving force, which, if exceeded, will enable the surface or interface to advance normal to itself, and, if not exceeded, will require the lateral growth mechanism.

Thus, for sufficiently large driving forces, the interface can move uniformly without the benefit of either a heterogeneous nucleation or screw dislocation mechanism. What constitutes a sufficiently large driving force depends upon the diffuseness of the interface, so that for extremely diffuse interfaces, this critical driving force will be so small that any measurable driving force will exceed it. Alternatively, for sharp interfaces, the critical driving force will be very large, and most growth will occur by the lateral step mechanism.

Note that in a typical solidification or crystallization process, the thermodynamic driving force is dictated by the degree of supercooling.

Morphology

Silver sulfide whiskers growing out of surface-mount resistors.

It is generally believed that the mechanical and other properties of the crystal are also pertinent to the subject matter, and that crystal morphology provides the missing link between growth kinetics and physical properties. The necessary thermodynamic apparatus was provided by Josiah Willard Gibbs›study of heterogeneous equilibrium. He provided a clear definition of surface energy, by which the concept of surface tension is made applicable to solids as well as liquids. He also appreciated that *an anisotropic surface free energy implied a non-spherical equilibrium shape*, which should be thermodynamically defined as *the shape which minimizes the total surface free energy*.

It may be instructional to note that whisker growth provides the link between the mechanical phenomenon of high strength in whiskers and the various growth mechanisms which are responsible for their fibrous morphologies. (Prior to the discovery of carbon nanotubes, single-crystal whiskers had the highest tensile strength of any materials known). Some mechanisms produce defect-free whiskers, while others may have single screw dislocations along the main axis of growth — producing high strength whiskers.

The mechanism behind whisker growth is not well understood, but seems to be encouraged by compressive mechanical stresses including mechanically induced stresses, stresses induced by diffusion of different elements, and thermally induced stresses. Metal whiskers differ from metallic dendrites in several respects. Dendrites are fern-shaped like the branches of a tree, and grow across the surface of the metal. In contrast, whiskers are fibrous and project at a right angle to the surface of growth, or substrate.

Diffusion-Control

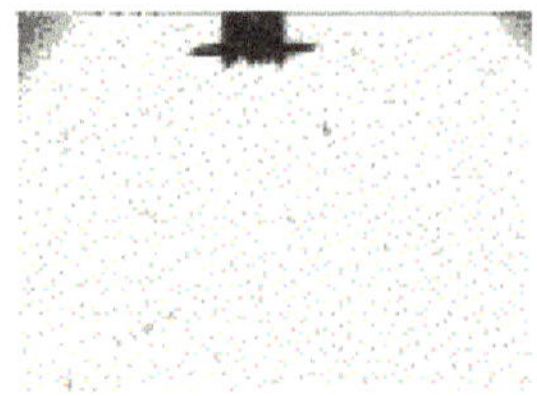

NASA animation of dendrite formation in microgravity.

Manganese dendrites on a limestone bedding plane from Solnhofen, Germany. Scale in mm.

Very commonly when the supersaturation (or degree of supercooling) is high, and sometimes even when it is not high, growth kinetics may be diffusion-controlled. Under such conditions, the polyhedral crystal form will be unstable, it will sprout protrusions at its corners and edges where the degree of supersaturation is at its highest level. The tips of these protrusions will clearly be the points of highest supersaturation. It is generally believed that the protrusion will become longer (and thinner at the tip) until the effect of interfacial free energy in raising the chemical potential slows the tip growth and maintains a constant value for the tip thickness.

In the subsequent tip-thickening process, there should be a corresponding instability of shape. Minor bumps or "bulges" should be exaggerated — and develop into rapidly growing side branches. In such an unstable (or metastable) situation, minor degrees of anisotropy should be sufficient to determine directions of significant branching and growth. The most appealing aspect of this argument, of course, is that it yields the primary morphological features of dendritic growth.

REFERENCES

1. Stephen Lower. "Chem1 online textbook—States of matter". Retrieved 2016-09-19.
2. Ashcroft and Mermin (1976). Solid state physics.
3. κρύσταλλος, Henry George Liddell, Robert Scott, A Greek-English Lexicon, on Perseus Digital Library
4. κρύος, Henry George Liddell, Robert Scott, A Greek-English Lexicon, on Perseus Digital Library
5. The American Heritage Dictionary of the English Language. Kreus. 2000.
6. Regal, Brian. (2009). Pseudoscience: A Critical Encyclopedia. Greenwood. p. 51. ISBN 978-0-313-35507-3
7. Patti Wigington (31 August 2016).Using Crystals and Gemstones in Magic. About.com. Retrieved 14 November 2016.
8. The Magic of Crystals and Gemstones. WitchesLore. 14 December 2011. Retrieved 14 November 2016.
9. The surface science of metal oxides, by Victor E. Henrich, P. A. Cox, page 28, google books link
10. G. Cressey and I. F. Mercer, (1999) Crystals, London, Natural History Museum, page 58
11. One or more of the preceding sentences incorporates text from a publication now in the public domain: Chisholm, Hugh, ed. (1911).Petrology. Encyclopædia Britannica (11th ed.). Cambridge University Press.
12. Cave of Crystal Giants — National Geographic Magazine. nationalgeographic.com.
13. International Union of Crystallography (1992).Report of the Executive Committee for 1991. Acta Crystallogr. A. 48 *(6): 922.* doi:10.1107/S0108767392008328.
14. Steurer W. (2004).Twenty years of structure research on quasicrystals. Part I. Pentagonal, octagonal, decagonal and dodecagonal quasicrystals. Z. Kristallogr. 219 *(7–2004): 391–446. Bibcode*:2004ZK....219..391S. doi:10.1524/zkri.219.7.391.35643.
15. The Nobel Prize in Chemistry 2011. Nobelprize.org. Retrieved 2011-12-29.

CHAPTER 2

X-ray Crystallography

Chapter Contents

- **Introduction**
- **History**
- **Contributions to Chemistry and Material Science**
- **Relationship to Other Scattering Techniques**
- **Methods**
- **Diffraction Theory**
- **Nobel Prizes for X-Ray Crystallography**
- **Microscopy**
- **Powder Diffraction**
- **Neutron Diffraction**

X-ray Crystallography is a scientific method used to determine the arrangement of atoms of a crystalline solid in three dimensional space. This technique takes advantage of the interatomic spacing of most crystalline solids by employing them as a diffraction gradient for x-ray light, which has wavelengths on the order of 1 angstrom (10^{-8} cm).

INTRODUCTION

X-ray crystallography is a technique used for determining the atomic and molecular structure of a crystal, in which the crystalline atoms cause a beam of incident X-rays to diffract into many specific directions. By measuring the angles and intensities of these diffracted beams, a crystallographer can produce a three-dimensional picture of the density of electrons within the crystal. From this electron density, the mean positions of the atoms in the crystal can be determined, as well as their chemical bonds, their disorder, and various other information.

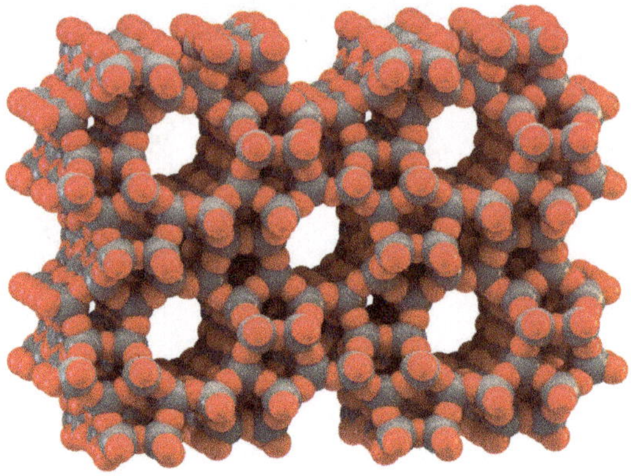

X-ray crystallography provides an atomistic model of zeolite, an aluminosilicate.

Since many materials can form crystals—such as salts, metals, minerals, semiconductors, as well as various inorganic, organic, and biological molecules—X-ray crystallography has been fundamental in the development of many scientific fields. In its first decades of use, this method determined the size of atoms, the lengths and types of chemical bonds, and the atomic-scale differences among various materials, especially minerals and alloys. The method also revealed the structure and function of many biological molecules, including vitamins, drugs, proteins and nucleic acids such as DNA. X-ray crystallography is still the chief method for characterizing the atomic structure of new materials and in discerning materials that appear similar by other experiments. X-ray crystal structures can also account for unusual electronic or elastic properties of a material, shed light on chemical interactions and processes, or serve as the basis for designing pharmaceuticals against diseases.

In a single-crystal X-ray diffraction measurement, a crystal is mounted on a goniometer. The goniometer is used to position the crystal at selected orientations. The crystal is illuminated with a finely focused monochromatic beam of X-rays, producing a diffraction pattern of regularly spaced spots known as reflections. The two-dimensional images taken at different orientations are converted into a three-dimensional model of the density of electrons within the crystal using the mathematical method of Fourier transforms, combined with chemical data known for the sample. Poor resolution (fuzziness) or even errors may result if the crystals are too small, or not uniform enough in their internal makeup.

X-ray crystallography is related to several other methods for determining atomic structures. Similar diffraction patterns can be produced by scattering electrons or neutrons, which are likewise interpreted by Fourier transformation. If single crystals of sufficient size cannot be obtained, various other X-ray methods can be applied to obtain less detailed information; such methods include fiber diffraction, powder diffraction and (if the sample is not crystallized) small-angle X-ray scattering (SAXS). If the material under investigation is only available in the form of nanocrystalline powders or suffers from poor crystallinity, the methods of electron crystallography can be applied for determining the atomic structure.

For all above mentioned X-ray diffraction methods, the scattering is elastic; the scattered X-rays have the same wavelength as the incoming X-ray. By contrast, inelastic X-ray scattering methods are useful in studying

HISTORY

Early Scientific History of Crystals and X-rays

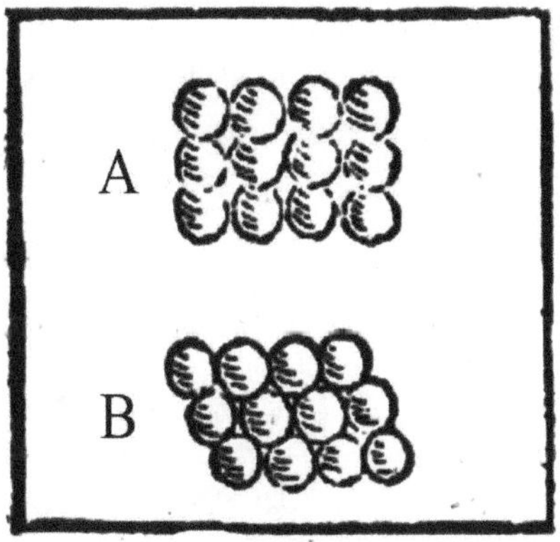

Drawing of square (Figure A, above) and hexagonal (Figure B, below) packing from Kepler's work, Strena seu de Nive Sexangula.

Crystals, though long admired for their regularity and symmetry, were not investigated scientifically until the 17th century. Johannes Kepler hypothesized in his work Strena seu de Nive Sexangula (A New Year's Gift of Hexagonal Snow) (1611) that the hexagonal symmetry of snowflake crystals was due to a regular packing of spherical water particles.

As shown by X-ray crystallography, the hexagonal symmetry of snowflakes results from the tetrahedral arrangement of hydrogen bonds about each water molecule. The water molecules are arranged similarly to the silicon atoms in the tridymitepolymorph of SiO_2. The resulting crystal structure has hexagonal symmetry when viewed along a principal axis.

The Danish scientist Nicolas Steno (1669) pioneered experimental investigations of crystal symmetry. Steno showed that the angles between the faces are the same in every exemplar of a particular type of crystal, and René Just Haüy (1784) discovered that every face of a crystal can be described by simple stacking patterns of blocks of the same shape and size. Hence, William Hallowes Miller in 1839 was able to give each face a unique label of three small integers, the Miller indices which remain in use today for identifying crystal faces. Haüy's study led to the correct idea that crystals are a regular three-dimensional array (a Bravais lattice) of atoms and molecules; a single unit cell is repeated indefinitely along three principal directions that are not necessarily perpendicular. In the 19th century, a complete catalog of the possible symmetries of a crystal was worked out by Johan Hessel, Auguste Bravais, Evgraf Fedorov,Arthur Schönflies and (belatedly) William Barlow (1894). From the available data and physical reasoning, Barlow proposed several crystal structures in the 1880s that were validated later by X-ray crystallography; however, the available data were too scarce in the 1880s to accept his models as conclusive.

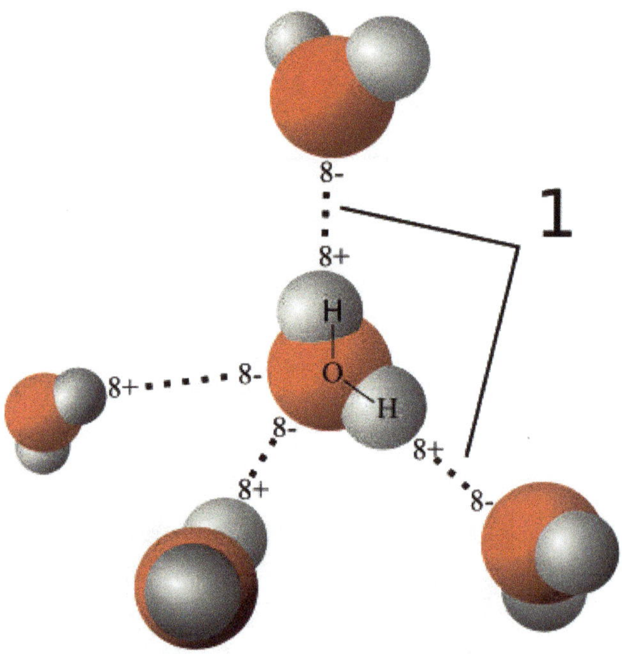

X-ray crystallography shows the arrangement of water molecules in ice, revealing the hydrogen bonds (1) that hold the solid together. Few other methods can determine the structure of matter with such precision (resolution).

Wilhelm Röntgen discovered X-rays in 1895, just as the studies of crystal symmetry were being concluded. Physicists were initially uncertain of the nature of X-rays, but soon suspected (correctly) that they were waves of electromagnetic radiation, in other words, another form of light. At that time, the wave model of light—specifically, the Maxwell theory of electromagnetic radiation—was well accepted among scientists, and experiments by Charles Glover Barkla showed that X-rays exhibited phenomena associated with electromagnetic waves, including transverse polarization and spectral lines akin to those observed in the visible wavelengths. Single-slit experiments in the laboratory of Arnold Sommerfeld suggested that X-rays had a wavelength of about 1 angstrom. However, X-rays are composed of photons, and thus are not only waves of electromagnetic radiation but also exhibit particle-like properties. Albert Einstein introduced the photon concept in 1905, but it was not broadly accepted until 1922, when Arthur Compton confirmed it by the scattering of X-rays from electrons. Therefore, these particle-like properties of X-rays, such as their ionization of gases, caused William Henry Bragg to argue in 1907 that X-rays were not electromagnetic radiation. Nevertheless, Bragg's view was not broadly accepted and the observation of X-ray diffraction by Max von Laue in 1912confirmed for most scientists that X-rays were a form of electromagnetic radiation.

X-ray Diffraction

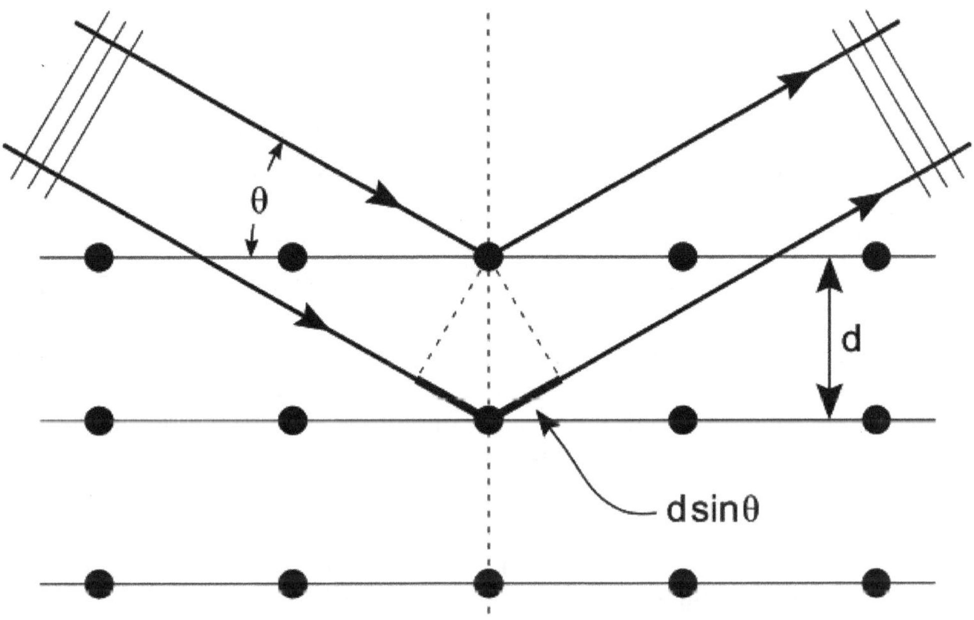

The incoming beam (coming from upper left) causes each scatterer to re-radiate a small portion of its intensity as a spherical wave. If scatterers are arranged symmetrically with a separation d, these spherical waves will be in sync (add constructively) only in directions where their path-length difference 2d sin θ equals an integer multiple of the wavelength λ. In that case, part of the incoming beam is deflected by an angle 2θ, producing a reflection spot in the diffraction pattern.

Crystals are regular arrays of atoms, and X-rays can be considered waves of electromagnetic radiation. Atoms scatter X-ray waves, primarily through the atoms' electrons. Just as an ocean wave striking a lighthouse produces secondary circular waves emanating from the lighthouse, so an X-ray striking an electron produces secondary spherical waves emanating from the electron. This phenomenon is known as elastic scattering, and the electron (or lighthouse) is known as the scatterer. A regular array of scatterers produces a regular array of spherical waves. Although these waves cancel one another out in most directions through destructive interference, they add constructively in a few specific directions, determined by Bragg's law:

$$2d \sin \theta = n\lambda$$

Here d is the spacing between diffracting planes, θ is the incident angle, n is any integer, and λ is the wavelength of the beam. These specific directions appear as spots on the diffraction pattern called reflections. Thus, X-ray diffraction results from an electromagnetic wave (the X-ray) impinging on a regular array of scatterers (the repeating arrangement of atoms within the crystal).

X-rays are used to produce the diffraction pattern because their wavelength λ is typically the same order of magnitude (1–100 angstroms) as the spacing d between planes in the crystal. In principle, any wave impinging on a regular array of scatterers produces diffraction, as predicted first by Francesco Maria Grimaldi in 1665. To produce significant diffraction, the spacing between the scatterers and the wavelength of the impinging wave should be similar in size. For illustration, the diffraction of sunlight through a bird's feather was first reported by James Gregory in the later 17th century. The first artificial diffraction gratings for visible light were constructed by David Rittenhouse in 1787, and Joseph von Fraunhofer in 1821. However, visible light has too long a wavelength (typically, 5500 angstroms) to observe diffraction from crystals. Prior to the first X-ray diffraction experiments, the spacings between lattice planes in a crystal were not known with certainty.

The idea that crystals could be used as a diffraction grating for X-rays arose in 1912 in a conversation between Paul Peter Ewald and Max von Laue in the English Garden in Munich. Ewald had proposed a resonator model of crystals for his thesis, but this model could not be validated using visible light, since the wavelength was much larger than the spacing between the resonators. Von Laue realized that electromagnetic radiation of a shorter wavelength was needed to observe such small spacings, and suggested that X-rays might have a wavelength comparable to the unit-cell spacing in crystals. Von Laue worked with two technicians, Walter Friedrich and his assistant Paul Knipping, to shine a beam of X-rays through a copper sulfate crystal and record its diffraction on a photographic plate. After being developed, the plate showed a large number of well-defined spots arranged in a pattern of intersecting circles around the spot produced by the central beam. Von Laue developed a law that connects the scattering angles and the size and orientation of the unit-cell spacings in the crystal, for which he was awarded the Nobel Prize in Physics in 1914.

Scattering

As described in the mathematical derivation below, the X-ray scattering is determined by the density of electrons within the crystal. Since the energy of an X-ray is much greater than that of a valence electron, the scattering may be modeled as Thomson scattering, the interaction of an electromagnetic ray with a free electron. This model is generally adopted to describe the polarization of the scattered radiation.

The intensity of Thomson scattering for one particle with mass m and charge q is:

$$I_o = I_e\left(\frac{q^4}{m^2c^4}\right)\frac{1+\cos^2 2\theta}{2} = I_e 7.94.10^{-26}\frac{1+\cos^2 2\theta}{2} = I_e f$$

Hence the atomic nuclei, which are much heavier than an electron, contribute negligibly to the scattered X-rays.

Development from 1912 to 1920

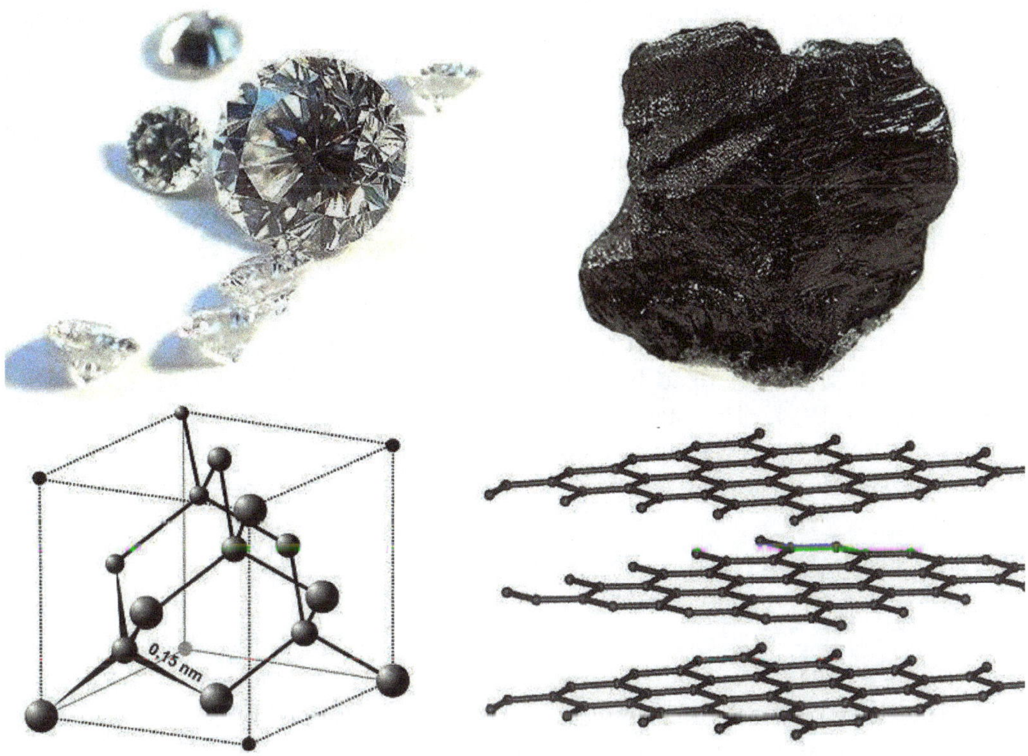

Although diamonds (top left) and graphite (top right) are identical in chemical composition—being both pure carbon—X-ray crystallography revealed the arrangement of their atoms (bottom) accounts for their different properties. In diamond, the carbon atoms are arranged tetrahedrally and held together by single covalent bonds, making it strong in all directions. By contrast, graphite is composed of stacked sheets. Within the sheet, the bonding is covalent and has hexagonal symmetry, but there are no covalent bonds between the sheets, making graphite easy to cleave into flakes.

After Von Laue's pioneering research, the field developed rapidly, most notably by physicists William Lawrence Bragg and his father William Henry Bragg. In 1912–1913, the younger Bragg developed Bragg's law, which connects the observed scattering with reflections from evenly spaced planes within the crystal. The Braggs, father and son, shared the 1915 Nobel Prize in Physics for their work in crystallography. The earliest structures were generally simple and marked by one-dimensional symmetry. However, as computational and experimental methods improved over the next decades, it became feasible to deduce reliable atomic positions for more complicated two- and three-dimensional arrangements of atoms in the unit-cell.

The potential of X-ray crystallography for determining the structure of molecules and minerals—then only known vaguely from chemical and hydrodynamic experiments—was realized immediately. The earliest structures were simple inorganic crystals and minerals, but even these revealed fundamental laws of physics and chemistry. The first atomic-resolution structure to besolved (i.e., determined) in 1914 was that of table salt. The distribution of electrons in the table-salt structure showed that crystals are not necessarily composed of covalently bonded molecules, and proved the existence of ionic compounds. The structure of diamond was solved in the same year, proving the tetrahedral arrangement of its chemical bonds and showing that the length of C–C single bond was 1.52 angstroms. Other early structures included copper, calcium fluoride (CaF_2, also known as fluorite), calcite ($CaCO_3$) and pyrite (FeS_2) in 1914; spinel ($MgAl_2O_4$) in 1915; the rutile and anatase forms of titanium dioxide (TiO_2) in 1916; pyrochroite $Mn(OH)_2$ and, by extension, brucite $Mg(OH)_2$ in 1919;.Also in 1919 sodium nitrate ($NaNO_3$) and caesium dichloroiodide ($CsICl_2$) were determined by Ralph Walter Graystone Wyckoff, and the wurtzite (hexagonal ZnS) structure became known in 1920.

The structure of graphite was solved in 1916 by the related method of powder diffraction, which was developed by Peter Debye and Paul Scherrer and, independently, by Albert Hull in 1917. The structure of graphite was determined from single-crystal diffraction in 1924 by two groups independently. Hull also used the powder method to determine the structures of various metals, such as iron and magnesium.

Cultural and Aesthetic Importance of X-ray Crystallography

In what has been called his scientific autobiography, The Development of X-ray Analysis, Sir William Lawrence Bragg mentioned that he believed the field of crystallography was particularly welcoming to women because the techno-aesthetics of the molecular structures resembled textiles and household objects. Bragg was known to compare crystal formation to "curtains, wallpapers, mosaics, and roses."

In 1951, the Festival Pattern Group at the Festival of Britain hosted a collaborative group of textile manufacturers and experienced crystallographers to design lace and prints based on the X-ray crystallography of insulin, china clay, and hemoglobin. One of the leading scientists of the project was Dr. Helen Megaw (1907–2002), the Assistant Director of Research at the Cavendish Laboratory in Cambridge at the time. Megaw is credited as one of the central figures who took inspiration from crystal diagrams and saw their potential in design. In 2008, the Wellcome Collection in London curated an exhibition on the Festival Pattern Group calledFrom Atom to Patterns.

CONTRIBUTIONS TO CHEMISTRY AND MATERIAL SCIENCE

X-ray crystallography has led to a better understanding of chemical bonds and non-covalent interactions. The initial studies revealed the typical radii of atoms, and confirmed many theoretical models of chemical bonding, such as the tetrahedral bonding of carbon in the diamond structure, the octahedral bonding of metals observed in ammonium hexachloroplatinate (IV), and the resonance observed in the planar carbonate group and in aromatic molecules. Kathleen Lonsdale's 1928 structure of hexamethylbenzene established the hexagonal symmetry of benzene and showed a clear difference in bond length between the aliphatic C–C bonds and aromatic C–C bonds; this finding led to the idea of resonance between chemical bonds, which had profound consequences for the development of chemistry. Her conclusions were anticipated by William Henry Bragg, who published models of naphthalene and anthracene in 1921 based on other molecules, an early form of molecular replacement.

Also in the 1920s, Victor Moritz Goldschmidt and later Linus Pauling developed rules for eliminating chemically unlikely structures and for determining the relative sizes of atoms. These rules led to the structure of brookite (1928) and an understanding of the relative stability of the rutile, brookite and anatase forms of titanium dioxide.

The distance between two bonded atoms is a sensitive measure of the bond strength and its bond order; thus, X-ray crystallographic studies have led to the discovery of even more exotic types of bonding in inorganic chemistry, such as metal-metal double bonds, metal-metal quadruple bonds, and three-center, two-electron bonds. X-ray crystallography—or, strictly speaking, an inelastic Compton scattering experiment—has also provided evidence for the partly covalent character of hydrogen bonds. In the field of organometallic chemistry, the X-ray structure of ferrocene initiated scientific studies of sandwich compounds, while that of Zeise's salt stimulated research into "back bonding" and metal-pi complexes. Finally, X-ray crystallography had a pioneering role in the development of supramolecular chemistry, particularly in clarifying the structures of the crown ethers and the principles of host-guest chemistry.

In material sciences, many complicated inorganic and organometallic systems have been analyzed using single-crystal methods, such as fullerenes, metalloporphyrins, and other complicated compounds. Single-crystal diffraction is also used in the pharmaceutical industry, due to recent problems with polymorphs. The major factors affecting the quality of single-crystal structures are the crystal's size and regularity; recrystallization is a commonly used technique to improve these factors in small-molecule crystals. The Cambridge Structural Database contains over 800,000 structures as of September 2016; over 99% of these structures were determined by X-ray diffraction.

Mineralogy and Metallurgy

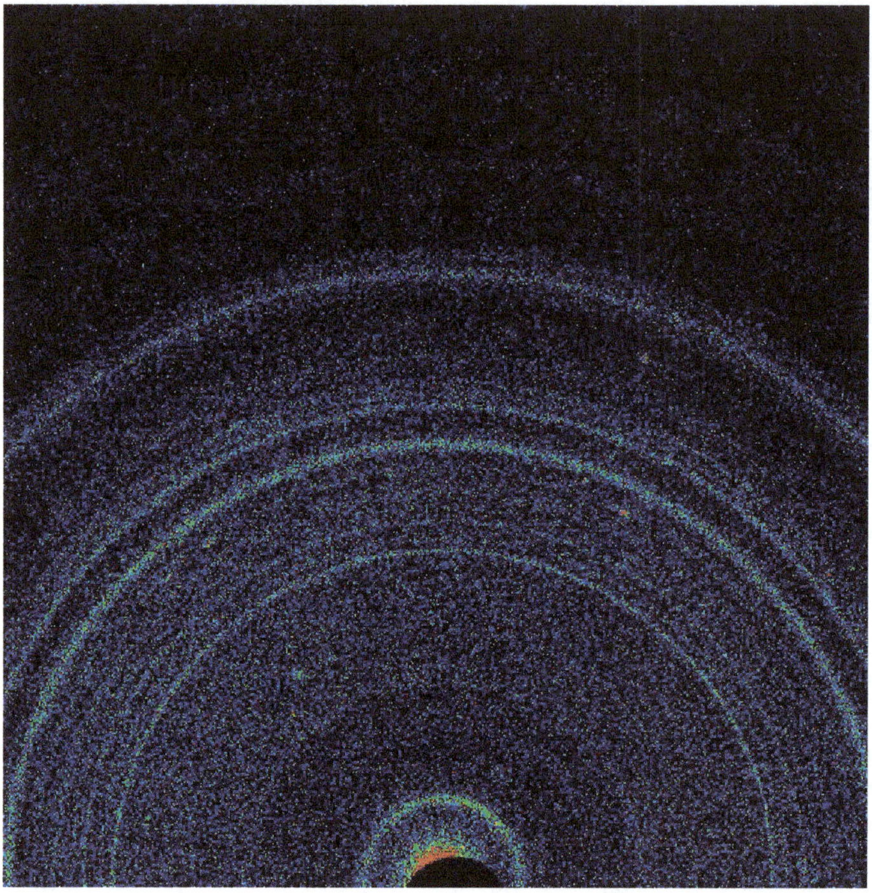

First X-ray diffraction view of Martian soil – CheMin analysis reveals feldspar, pyroxenes, olivine and more (Curiosity rover at "Rocknest, October 17, 2012).

Since the 1920s, X-ray diffraction has been the principal method for determining the arrangement of atoms in minerals and metals. The application of X-ray crystallography to mineralogy began with the structure of garnet, which was determined in 1924 by Menzer. A systematic X-ray crystallographic study of the silicates was undertaken in the 1920s. This study showed that, as the Si/O ratio is altered, the silicate crystals exhibit significant changes in their atomic arrangements. Machatschki extended these insights to minerals in which aluminium substitutes for the silicon atoms of the silicates. The first application of X-ray crystallography to metallurgy likewise occurred in the mid-1920s. Most notably, Linus Pauling's structure of the alloy Mg_2Sn led to his theory of the stability and structure of complex ionic crystals.

On October 17, 2012, the Curiosity rover on the planet Mars at "Rocknest performed the first X-ray diffraction analysis of Martian soil. The results from the rover's CheMin analyzer revealed the presence of several minerals, including feldspar, pyroxenes and olivine, and suggested that the Martian soil in the sample was similar to theweathered basaltic soils of Hawaiian volcanoes.

Early organic and small biological molecules

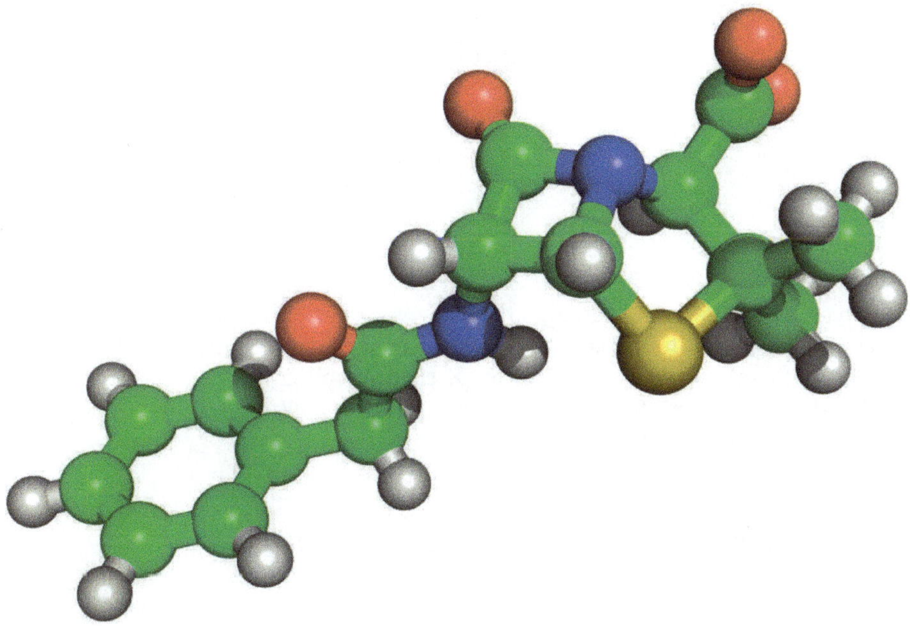

The three-dimensional structure of penicillin, solved by Dorothy Crowfoot Hodgkin in 1945. The green, white, red, yellow and blue spheres represent atoms of carbon, hydrogen, oxygen, sulfur and nitrogen, respectively.

The first structure of an organic compound, hexamethylenetetramine, was solved in 1923. This was followed by several studies of long-chain fatty acids, which are an important component of biological membranes. In the 1930s, the structures of much larger molecules with two-dimensional complexity began to be solved. A significant advance was the structure of phthalocyanine, a large planar molecule that is closely related to porphyrin molecules important in biology, such as heme, corrin and chlorophyll.

X-ray crystallography of biological molecules took off with Dorothy Crowfoot Hodgkin, who solved the structures of cholesterol (1937), penicillin (1946) and vitamin B12 (1956), for which she was awarded the Nobel Prize in Chemistry in 1964. In 1969, she succeeded in solving the structure of insulin, on which she worked for over thirty years.

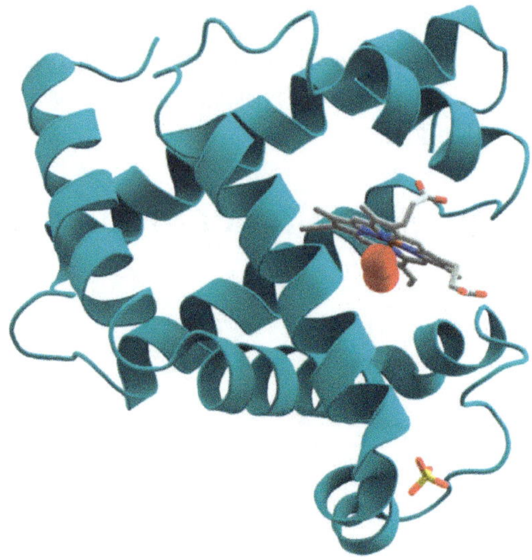

Ribbon diagram of the structure of myoglobin, showing colored alpha helices. Such proteins are long, linear molecules with thousands of atoms; yet the relative position of each atom has been determined with sub-atomic resolution by X-ray crystallography. Since it is difficult to visualize all the atoms at once, the ribbon shows the rough path of the protein polymer from its N-terminus (blue) to its C-terminus (red).

Biological Macromolecular Crystallography

Crystal structures of proteins (which are irregular and hundreds of times larger than cholesterol) began to be solved in the late 1950s, beginning with the structure of sperm whale myoglobin by Sir John Cowdery Kendrew, for which he shared the Nobel Prize in Chemistry with Max Perutz in 1962. Since that success, over 86817 X-ray crystal structures of proteins, nucleic acids and other biological molecules have been determined. For comparison, the nearest competing method in terms of structures analyzed is nuclear magnetic resonance (NMR) spectroscopy, which has resolved 9561 chemical structures. Moreover, crystallography can solve structures of arbitrarily large molecules, whereas solution-state NMR is restricted to relatively small ones (less than 70 kDa). X-ray crystallography is now used routinely by scientists to determine how a pharmaceutical drug interacts with its protein target and what changes might improve it. However, intrinsic membrane proteins remain challenging to crystallize because they require detergents or other means to solubilize them in isolation, and such detergents often interfere with crystallization. Such membrane proteins are a large component of the genome, and include many proteins of great physiological importance, such as ion channels and receptors. Helium cryogenics are used to prevent radiation damage in protein crystals.

On the other end of the size scale, relative small molecules are able to lure the resolving power of X-ray crystallography. The structure assigned in 1991 to the antibiotic isolated from a marina organism, diazonamide A – $C_{40}H_{34}Cl_2N_6O_6$ with M = 765.65 –

proved to be incorrect by the classical proof of structure: a synthetic sample was not identical to the natural product. The mistake was possible because of the inability of X-ray crystallography to distinguish between the correct -OH / >NH and the interchanged -NH_2 / -O- groups in the incorrect structure.

RELATIONSHIP TO OTHER SCATTERING TECHNIQUES

Elastic vs. Inelastic Scattering

X-ray crystallography is a form of elastic scattering; the outgoing X-rays have the same energy, and thus same wavelength, as the incoming X-rays, only with altered direction. By contrast, inelastic scattering occurs when energy is transferred from the incoming X-ray to the crystal, e.g., by exciting an inner-shell electron to a higher energy level. Such inelastic scattering reduces the energy (or increases the wavelength) of the outgoing beam. Inelastic scattering is useful for probing such excitations of matter, but not in determining the distribution of scatterers within the matter, which is the goal of X-ray crystallography.

X-rays range in wavelength from 10 to 0.01 nanometers; a typical wavelength used for crystallography is 1 Å (0.1 nm), which is on the scale of covalent chemical bonds and the radius of a single atom. Longer-wavelength photons (such as ultraviolet radiation) would not have sufficient resolution to determine the atomic positions. At the other extreme, shorter-wavelength photons such as gamma rays are difficult to produce in large numbers, difficult to focus, and interact too strongly with matter, producing particle-antiparticle pairs. Therefore, X-rays are thesweetspot for wavelength when determining atomic-resolution structures from the scattering of electromagnetic radiation.

Other X-ray Techniques

Other forms of elastic X-ray scattering include powder diffraction, Small-Angle X-ray Scattering (SAXS) and several types of X-ray fiber diffraction, which was used by Rosalind Franklin in determining the double-helix structure of DNA. In general, single-crystal X-ray diffraction offers more structural information than these other techniques; however, it requires a sufficiently large and regular crystal, which is not always available.

These scattering methods generally use monochromatic X-rays, which are restricted to a single wavelength with minor deviations. A broad spectrum of X-rays (that is, a blend of X-rays with different wavelengths) can also be used to carry out X-ray diffraction, a technique known as the Laue method. This is the method used in the original discovery of X-ray diffraction. Laue scattering provides much structural information with only a short exposure to the X-ray beam, and is therefore used in structural studies of very rapid events (Time resolved crystallography). However, it is not as well-suited as monochromatic scattering for determining the full atomic structure of a crystal and therefore works better with crystals with relatively simple atomic arrangements.

The Laue back reflection mode records X-rays scattered backwards from a broad spectrum source. This is useful if the sample is too thick for X-rays to transmit through it. The diffracting planes in the crystal are determined by knowing that the normal to the diffracting plane bisects the angle between the incident beam and the diffracted beam. A Greninger chart can be usedto interpret the back reflection Laue photograph.

Electron and Neutron Diffraction

Other particles, such as electrons and neutrons, may be used to produce a diffraction pattern. Although electron, neutron, and X-ray scattering are based on different physical processes, the resulting diffraction patterns are analyzed using the same coherent diffraction imaging techniques.

As derived below, the electron density within the crystal and the diffraction patterns are related by a simple mathematical method, the Fourier transform, which allows the density to be calculated relatively easily from the patterns. However, this works only if the scattering is weak, i.e., if the scattered beams are much less intense than the incoming beam. Weakly scattered beams pass through the remainder of the crystal without undergoing a second scattering event. Such re-scattered waves are called "secondary scattering" and hinder the analysis. Any sufficiently thick crystal will produce secondary scattering, but since X-rays interact relatively weakly with the electrons, this is generally not a significant concern. By contrast, electron beams may produce strong secondary scattering even for relatively thin crystals (>100 nm). Since this thickness corresponds to the diameter of many viruses, a promising direction is the electron diffraction of isolated macromolecular assemblies, such as viral capsids and molecular machines, which may be carried out with a cryo-electron microscope. Moreover, the strong interaction of electrons with matter (about 1000 times stronger than for X-rays) allows determination of the atomic structure of extremely small volumes. The field of applications for electron crystallography ranges from bio molecules like membrane proteins over organic thin films to the complex structures of (nanocrystalline) intermetallic compounds and zeolites.

Neutron diffraction is an excellent method for structure determination, although it has been difficult to obtain intense, monochromatic beams of neutrons in sufficient quantities. Traditionally, nuclear reactors have been used, although sources producing neutrons by spallation are becoming increasingly available. Being uncharged, neutrons scatter much more readily from the atomic nuclei rather than from the electrons. Therefore, neutron scattering is very useful for observing the positions of light atoms with few electrons, especially hydrogen, which is essentially invisible in the X-ray diffraction. Neutron scattering also has the remarkable property that the solvent can be made invisible by adjusting the ratio of normal water, H_2O, and heavy water, D_2O.

METHODS

Overview of single-crystal X-ray Diffraction

The oldest and most precise method of X-ray crystallography is single-crystal X-ray diffraction, in which a beam of X-rays strikes a single crystal, producing scattered beams. When they land on a piece of film or other detector, these beams make a diffraction pattern of spots; the strengths and angles of these beams are recorded as the crystal is gradually rotated. Each spot is called a reflection, since it corresponds to the reflection of the X-rays from one set of evenly spaced planes within the crystal. For single crystals of sufficient purity and regularity, X-ray diffraction data can determine the mean chemical bond lengths and angles to within a few thousandths of an angstrom and to within a few tenths of a degree, respectively. The atoms in a crystal are not static, but oscillate about their mean positions, usually by less than a few tenths of an angstrom. X-ray crystallography allows measuring the size of these oscillations.

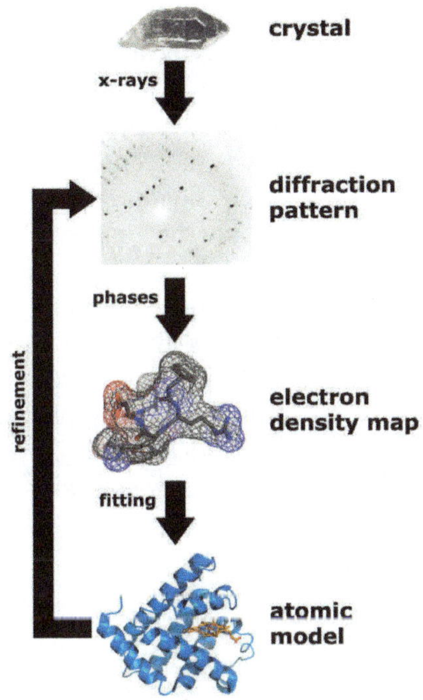

Workflow for solving the structure of a molecule by X-ray crystallography.

Procedure

The technique of single-crystal X-ray crystallography has three basic steps. The first—and often most difficult—step is to obtain an adequate crystal of the material under study. The crystal should be sufficiently large (typically larger than 0.1 mm in all dimensions), pure in composition and regular in structure, with no significant internal imperfections such as cracks or twinning.

In the second step, the crystal is placed in an intense beam of X-rays, usually of a single wavelength (monochromatic X-rays), producing the regular pattern of reflections. As the crystal is gradually rotated, previous reflections disappear and new ones appear; the intensity of every spot is recorded at every orientation of the crystal. Multiple data sets may have to be collected, with each set covering slightly more than half a full rotation of the crystal and typically containing tens of thousands of reflections.

In the third step, these data are combined computationally with complementary chemical information to produce and refine a model of the arrangement of atoms within the crystal. The final, refined model of the atomic arrangement—now called a crystal structure—is usually stored in a public database.

Limitations

As the crystal's repeating unit, its unit cell, becomes larger and more complex, the atomic-level picture provided by X-ray crystallography becomes less well-resolved (more "fuzzy") for a given number of observed reflections. Two limiting cases of X-ray crystallography—"small-molecule" (which includes continuous inorganic solids) and "macromolecular" crystallography—are often discerned. Small-molecule crystallography typically involves crystals with fewer than 100 atoms in their asymmetric unit; such crystal structures are usually so well resolved that the atoms can be discerned as isolatedblobs of electron density. By contrast, macromolecular crystallography often involves tens of thousands of atoms in the unit cell. Such crystal structures are generally less well-resolved (more "smeared out"); the atoms and chemical bonds appear as tubes of electron density, rather than as isolated atoms. In general, small molecules are also easier to crystallize than macromolecules; however, X-ray crystallography has proven possible even for viruses with hundreds of thousands of atoms. Though normally x-ray crystallography can only be performed if the sample is in crystal form, new research has been done into sampling non-crystalline forms of samples.

Crystallization

A protein crystal seen under a microscope. Crystals used in X-ray crystallography may be smaller than a millimeter across.

Although crystallography can be used to characterize the disorder in an impure or irregular crystal, crystallography generally requires a pure crystal of high regularity to solve the structure of a complicated arrangement of atoms. Pure, regular crystals can sometimes be obtained from natural or synthetic materials, such as samples of metals, minerals or other macroscopic materials. The regularity of such crystals can sometimes be improved with macromolecular crystal annealing and other methods. However, in many cases, obtaining a diffraction-quality crystal is the chief barrier to solving its atomic-resolution structure.

Small-molecule and macromolecular crystallography differ in the range of possible techniques used to produce diffraction-quality crystals. Small molecules generally have few degrees of conformational freedom, and may be crystallized by a wide range of methods, such as chemical vapor deposition and recrystallization. By contrast, macromolecules generally have many degrees of freedom and their crystallization must be carried out so as to maintain a stable structure. For example, proteins and larger RNA molecules cannot be crystallized if their tertiary structure has been unfolded; therefore, the range of crystallization conditions is restricted to solution conditions in which such molecules remain folded.

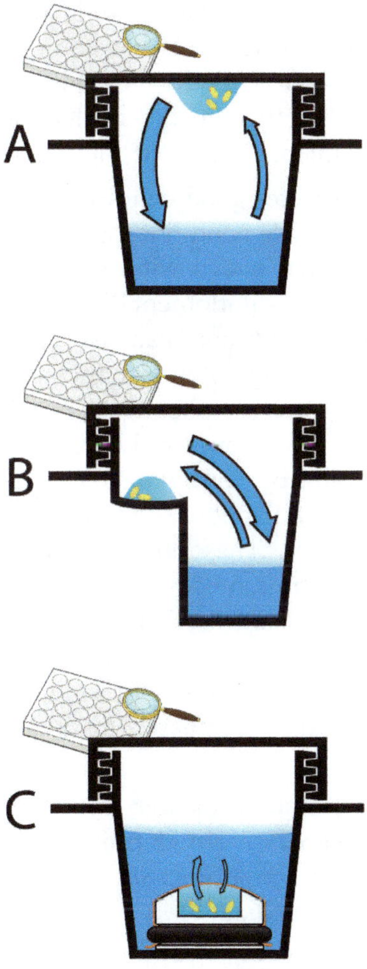

Three methods of preparing crystals, A: Hanging drop. B: Sitting drop. C: Microdialysis.

Protein crystals are almost always grown in solution. The most common approach is to lower the solubility of its component molecules very gradually; if this is done too quickly, the molecules will precipitate from solution, forming a useless dust or amorphous gel on the bottom of the container. Crystal growth in solution is characterized by two steps: nucleation of a microscopic crystallite (possibly having only 100 molecules), followed by growth of that crystallite, ideally to a diffraction-quality crystal. The solution conditions that favor the first step (nucleation) are not always the same conditions that favor the second step (subsequent growth). The crystallographer's goal is to identify solution conditions that favor the development of a single, large crystal, since larger crystals offer improved resolution of the molecule. Consequently, the solution conditions should disfavor the first step (nucleation) but favor the second (growth), so that only one large crystal forms per droplet. If nucleation is favored too much, a shower of small crystallites will form in the droplet, rather than one large crystal; if favored too little, no crystal will form whatsoever. Other approaches involves, crystallizing proteins under oil, where aqueous protein solutions are dispensed under liquid oil, and water evaporates through the layer of oil. Different oils have different evaporation permeabilities, therefore yielding changes in concentration rates from different percipient/protein mixture. The technique relies on bringing the protein directly into the nucleation zone by mixing protein with the appropriate amount of percipient to prevent the diffusion of water out of the drop.

It is extremely difficult to predict good conditions for nucleation or growth of well-ordered crystals. In practice, favorable conditions are identified by screening; a very large batch of the molecules is prepared, and a wide variety of crystallization solutions are tested.Hundreds, even thousands, of solution conditions are generally tried before finding the successful one. The various conditions can use one or more physical mechanisms to lower the solubility of the molecule; for example, some may change the pH, some contain salts of the Hofmeister series or chemicals that lower the dielectric constant of the solution, and still others contain large polymers such as polyethylene glycol that drive the molecule out of solution by entropic effects. It is also common to try several temperatures for encouraging crystallization, or to gradually lower the temperature so that the solution becomes supersaturated. These methods require large amounts of the target molecule, as they use high concentration of the molecule(s) to be crystallized. Due to the difficulty in obtaining such large quantities (milligrams) of crystallization-grade protein, robots have been developed that are capable of accurately dispensing crystallization trial drops that are in the order of 100 nanoliters in volume. This means that 10-fold less protein is used per experiment when compared to crystallization trials set up by hand (in the order of 1 microliter).

Several factors are known to inhibit or mar crystallization. The growing crystals are generally held at a constant temperature and protected from shocks or vibrations that might disturb their crystallization. Impurities in the molecules or in the crystallization solutions are often inimical to crystallization. Conformational flexibility in the molecule also tends to make crystallization less likely, due to entropy. Ironically, molecules that tend to self-assemble into regular helices are often unwilling to assemble into crystals. Crystals can be marred by twinning, which can occur when a unit cell can pack equally favorably in multiple orientations; although recent advances in computational methods may allow solving the structure of some twinned crystals. Having failed to crystallize a target molecule, a crystallographer may try again with a slightly modified version of the molecule; even small changes in molecular properties can lead to large differences in crystallization behavior.

Data Collection

Mounting the Crystal

The crystal is mounted for measurements so that it may be held in the X-ray beam and rotated. There are several methods of mounting. In the past, crystals were loaded into glass capillaries with the crystallization solution (the mother liquor). Nowadays, crystals of small molecules are typically attached with oil or glue to a glass fiber or a loop, which is made of nylon or plastic and attached to a solid rod. Protein crystals are scooped up by a loop, then flash-frozen with liquid nitrogen. This freezing reduces the radiation damage of the X-rays, as well as the noise in the Bragg peaks due to thermal motion (the Debye-Waller effect). However, untreated protein crystals often crack if flash-frozen; therefore, they are generally pre-soaked in a cryoprotectant solution before freezing. Unfortunately, this pre-soak may itself cause the crystal to crack, ruining it for crystallography. Generally, successful cryo-conditions are identified by trial and error.

The capillary or loop is mounted on a goniometer, which allows it to be positioned accurately within the X-ray beam and rotated. Since both the crystal and the beam are often very small, the crystal must be centered within the beam to within ~25 micrometers accuracy, which is aided by a camera focused on the crystal. The most common type of goniometer is the "kappa goniometer", which offers three angles of rotation: the ω angle, which rotates about an axis perpendicular to the beam; the κ angle, about an axis at ~50° to the ω axis; and, finally, the φ angle about the loop/capillary axis. When the κ angle is zero, the ω and φ axes are aligned. The κ rotation allows for convenient mounting of the crystal, since the arm in which the crystal is mounted may be swung out towards the crystallographer. The oscillations carried out during data collection (mentioned below) involve the ω axis only. An older type of goniometer is the four-circle goniometer, and its relatives such as the six-circle goniometer.

X-ray Sources

Rotating Anode

Small scale can be done on a local X-ray tube source, typically coupled with an image plate detector. These have the advantage of being (relatively) inexpensive and easy to maintain, and allow for quick screening and collection of samples. However, the wavelength light produced is limited by anode material, typically copper. Further, intensity is limited by the power applied and cooling capacity available to avoid melting the anode. In such systems, electrons are boiled off of a cathode and accelerated through a strong electric potential of ~50 kV; having reached a high speed, the electrons collide with a metal plate, emitting bremsstrahlung and some strong spectral lines corresponding to the excitation of inner-shell electrons of the metal. The most common metal used is copper, which can be kept cool easily, due to its high thermal conductivity, and which produces strong K_α and K_β lines. The K_β line is sometimes suppressed with a thin (~10 μm) nickel foil. The simplest and cheapest variety of sealed X-ray tube has a stationary anode (the Crookes tube) and run with ~2 kW of electron beam power. The more expensive variety has a rotating-anode type source that run with ~14 kW of e-beam power.

X-rays are generally filtered (by use of X-Ray Filters) to a single wavelength (made monochromatic) and collimated to a single direction before they are allowed to strike the crystal. The filtering not only simplifies the data analysis, but also removes radiation that degrades the crystal without contributing useful information. Collimation is done either with a collimator (basically, a long tube) or with a clever arrangement of gently curved mirrors. Mirror systems are preferred for small crystals (under 0.3 mm) or with large unit cells (over 150 Å)

Synchrotron Radiation

Synchrotron radiation are some of the brightest lights on earth. It is the single most powerful tool available to X-ray crystallographers. It is made of X-ray beams generated in large machines called synchrotrons. These machines accelerate electrically charged particles, often electrons, to nearly the speed of light and confine them in a (roughly) circular loop using magnetic fields.

Synchrotrons are generally national facilities, each with several dedicated beamlines where data is collected without interruption. Synchrotrons were originally designed for use by high-energy physicists studying subatomic particles and cosmic phenomena. The largest component of each synchrotron is its electron storage ring. This ring is actually not a perfect circle, but a many-sided polygon. At each corner of the polygon, or sector, precisely aligned magnets bend the electron stream. As the electrons' path is bent, they emit bursts of energy in the form of X-rays.

Using synchrotron radiation frequently has specific requirements for X-ray crystallography. The intense ionizing radiation can cause radiation damage to samples, particularly macromolecular crystals. Cryo crystallography protects the sample from radiation damage, by freezing the crystal at liquid nitrogen temperatures (~100 K). However, synchrotron radiation frequently has the advantage of user selectable wavelengths, allowing for anomalous scattering experiments which maximizes anomalous signal. This is critical in experiments such as SAD and MAD.

Free electron Laser

Recently, free electron lasers have been developed for use in X-ray crystallography. These are the brightest X-ray sources currently available; with the X-rays coming in femtosecond bursts. The intensity of the source is such that atomic resolution diffraction patterns can be resolved for crystals otherwise too small for collection. However, the intense light source also destroys the sample, requiring multiple crystals to be shot. As each crystal is randomly oriented in the beam, hundreds of thousands of individual diffraction images must be collected in order to get a complete data-set. This method, serial femtosecond crystallography, has been used in solving the structure of a number of protein crystal structures, sometimes noting differences with equivalent structures collected from synchrotron sources.

Recording the Reflections

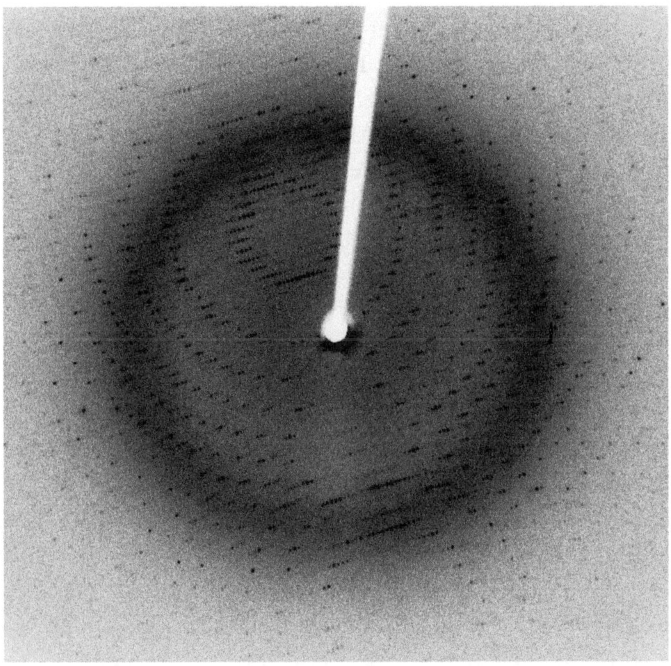

An X-ray diffraction pattern of a crystallized enzyme. The pattern of spots (reflections) and the relative strength of each spot (intensities) can be used to determine the structure of the enzyme.

When a crystal is mounted and exposed to an intense beam of X-rays, it scatters the X-rays into a pattern of spots or reflections that can be observed on a screen behind the crystal. A similar pattern may be seen by shining a laser pointer at a compact disc. The relative intensities of these spots provide the information to determine the arrangement of molecules within the crystal in atomic detail. The intensities of these reflections may be recorded with photographic film, an area detector or with a charge-coupled device (CCD) image sensor. The peaks at small angles correspond to low-resolution data, whereas those at high angles represent high-resolution data; thus, an upper limit on the eventual resolution of the structure can be determined from the first few images. Some measures of diffraction quality can be determined at this point, such as the mosaicity of the crystal and its overall disorder, as observed in the peak widths. Some pathologies of the crystal that would render it unfit for solving the structure can also be diagnosed quickly at this point.

One image of spots is insufficient to reconstruct the whole crystal; it represents only a small slice of the full Fourier transform. To collect all the necessary information, the crystal must be rotated step-by-step through 180°, with an image recorded at every step; actually, slightly more than 180° is required to cover reciprocal space, due to the curvature of the Ewald sphere. However, if the crystal has a higher symmetry, a smaller angular range such as 90° or 45° may be recorded. The rotation axis should be changed at least once, to avoid developing ablind spot in reciprocal space close to the rotation axis. It is customary to rock the crystal slightly (by 0.5–2°) to catch a broader region of reciprocal space.

Multiple data sets may be necessary for certain phasing methods. For example, MAD phasing requires that the scattering be recorded at least three (and usually four, for redundancy) wavelengths of the incoming X-ray radiation. A single crystal may degrade too much during the collection of one data set, owing to radiation damage; in such cases, data sets on multiple crystals must be taken.

Data analysis

Crystal Symmetry, Unit Cell, and Image Scaling

The recorded series of two-dimensional diffraction patterns, each corresponding to a different crystal orientation, is converted into a three-dimensional model of the electron density; the conversion uses the mathematical technique of Fourier transforms, which is explained below. Each spot corresponds to a different type of variation in the electron density; the crystallographer must determine which variation corresponds to which spot (indexing), the relative strengths of the spots in different images (merging and scaling) and how the variations should be combined to yield the total electron density (phasing).

Data processing begins with indexing the reflections. This means identifying the dimensions of the unit cell and which image peak corresponds to which position in reciprocal space. A byproduct of indexing is to determine the symmetry of the crystal, i.e., its space group. Some space groups can be eliminated from the beginning. For example, reflection symmetries cannot be observed in chiral molecules; thus, only 65 space groups of 230 possible are allowed for protein molecules which are almost always chiral. Indexing is generally accomplished using an autoindexing routine. Having assigned symmetry, the data is then integrated. This converts the hundreds of images containing the thousands of reflections into a single file, consisting of (at the very least) records of the Miller index of each reflection, and an intensity for each reflection (at this state the file often also includes error estimates and measures of partiality (what part of a given reflection was recorded on that image)).

A full data set may consist of hundreds of separate images taken at different orientations of the crystal. The first step is to merge and scale these various images, that is, to identify which peaks appear in two or more images (merging) and to scale the relative images so that they have a consistent intensity scale. Optimizing the intensity scale is critical because the relative intensity of the peaks is the key information from which the structure is determined. The repetitive technique of crystallographic data collection and the often high symmetry of crystalline materials cause the diffractometer to record many symmetry-equivalent reflections multiple times. This allows calculating the symmetry-related R-factor, a reliability index based upon how similar are the measured intensities of symmetry-equivalent reflections thus assessing the quality of the data.

Initial phAsing

The data collected from a diffraction experiment is a reciprocal space representation of the crystal lattice. The position of each diffraction 'spot' is governed by the size and shape of the unit cell, and the inherent symmetry within the crystal. The intensity of each diffraction 'spot' is recorded, and this intensity is proportional to the square of the structure factor amplitude. The structure factor is a complex number containing information relating to both the amplitude and phase of a wave. In order to obtain an interpretable electron density map, both amplitude and phase must be known (an electron density map allows a crystallographer to build a starting model of the molecule). The phase cannot be directly recorded during a diffraction experiment: this is known as the phase problem. Initial phase estimates can be obtained in a variety of ways:

- ***Ab initio* phasing** or **direct methods** – This is usually the method of choice for small molecules (<1000 non-hydrogen atoms), and has been used successfully to solve the phase problems for small proteins. If the resolution of the data is better than 1.4 Å (140 pm), direct methods can be used to obtain phase information, by exploiting known phase relationships between certain groups of reflections.
- **Molecular replacement** – if a related structure is known, it can be used as a search model in molecular replacement to determine the orientation and position of the molecules within the unit cell. The phases obtained this way can be used to generate electron density maps.
- **Anomalous X-ray scattering** (MAD or SAD phasing) – the X-ray wavelength may be scanned past an absorption edge of an atom, which changes the scattering in a known way. By recording full sets of reflections at three different wavelengths (far below, far above and in the middle of the absorption edge) one can solve for the substructure of the anomalously diffracting atoms and hence the structure of the whole molecule. The most popular method of incorporating anomalous scattering atoms into proteins is to express the protein in a methionine auxotroph (a host incapable of synthesizing methionine) in a media rich in seleno-methionine, which contains selenium atoms. A MAD experiment can then be conducted around the absorption edge, which should then yield the position of any methionine residues within the protein, providing initial phases.
- **Heavy atom methods** (multiple isomorphous replacement) – If electron-dense metal atoms can be introduced into the crystal, direct methods or Patterson-space methods can be used to determine their location and to obtain initial phases. Such heavy atoms can be introduced either by soaking the crystal in a heavy atom-containing solution, or by co-crystallization (growing the crystals in the presence of a heavy atom). As in MAD phasing, the changes in the scattering amplitudes can be interpreted to yield the phases. Although this is the original method by which protein crystal structures were solved, it has largely been superseded by MAD phasing with selenomethionine.

Model Building and Phase Refinement

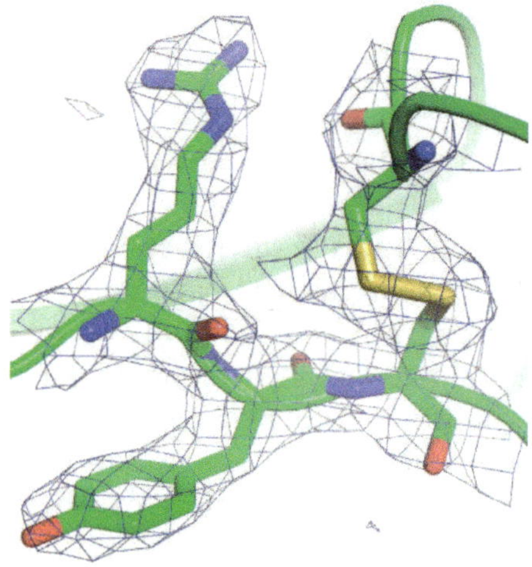

A protein crystal structure at 2.7 Å resolution. The mesh encloses the region in which the electron density exceeds a given threshold. The straight segments represent chemical bonds between the non-hydrogen atoms of an arginine (upper left), a tyrosine (lower left), a disulfide bond (upper right, in yellow), and some peptide groups (running left-right in the middle). The two curved green tubes represent spline fits to the polypeptide backbone.

Having obtained initial phases, an initial model can be built. This model can be used to refine the phases, leading to an improved model, and so on. Given a model of some atomic positions, these positions and their respective Debye-Waller factors (or **B**-factors, accounting for the thermal motion of the atom) can be refined to fit the observed diffraction data, ideally yielding a better set of phases. A new model can then be fit to the new electron density map and a further round of refinement is carried out. This continues until the correlation between the diffraction data and the model is maximized. The agreement is measured by an R-factor defined as

$$R = \frac{\sum_{\text{all reflections}} |F_o - F_c|}{\sum_{\text{all reflections}} |F_o|}$$

where F is the structure factor. A similar quality criterion is R_{free}, which is calculated from a subset (~10%) of reflections that were not included in the structure refinement. Both R factors depend on the resolution of the data. As a rule of thumb, R_{free} should be approximately the resolution in angstroms divided by 10; thus, a data-set with 2 Å resolution should yield a final $R_{free} \sim 0.2$. Chemical bonding features such as stereochemistry, hydrogen bonding and distribution of bond lengths and angles are complementary measures of the model quality. Phase bias is a serious problem in such iterative model building. Omit maps are a common technique used to check for this.

It may not be possible to observe every atom in the asymmetric unit. In many cases, disorder smears the electron density map. Weakly scattering atoms such as hydrogen are routinely invisible. It is also possible for a single atom to appear multiple times in an electron density map, e.g., if a protein sidechain has multiple (<4) allowed conformations. In still other cases, the crystallographer may detect that the covalent structure deduced for the molecule was incorrect, or changed. For example, proteins may be cleaved or undergo post-translational modifications that were not detected prior to the crystallization.

Disorder

A common challenge in refinement of crystal structures results from crystallographic disorder. Disorder can take many forms but in general involves the coexistence of two or more species or conformations. Failure to recognize disorder results in flawed interpretation. Pitfalls from improper modeling of disorder are illustrated by the discounted hypothesis of bond stretch isomerism. Disorder is modelled with respect to the relative population of the components, often only two, and their identity. In structures of large molecules and ions, solvent and counterions are often disordered.

Deposition of the Structure

Once the model of a molecule's structure has been finalized, it is often deposited in a crystallographic database such as the Cambridge Structural Database (for small molecules), the Inorganic Crystal Structure Database (ICSD) (for inorganic compounds) or the Protein Data Bank (for protein structures). Many structures obtained in private commercial ventures to crystallize medicinally relevant proteins are not deposited in public crystallographic databases.

DIFFRACTION THEORY

The main goal of X-ray crystallography is to determine the density of electrons f(**r**) throughout the crystal, where **r** represents the three-dimensional position vector within the crystal. To do this, X-ray scattering is used to collect data about its Fourier transform F(**q**), which is inverted mathematically to obtain the density defined in real space, using the formula

$$f(\mathbf{r}) = \frac{1}{(2\pi)^3} \int F(\mathbf{q}) e^{i\mathbf{q}\cdot\mathbf{r}} d\mathbf{q}$$

where the integral is taken over all values of **q**. The three-dimensional real vector **q** represents a point in reciprocal space, that is, to a particular oscillation in the electron density as one moves in the direction in which **q** points. The length of **q** corresponds to 2π divided by the wavelength of the oscillation. The corresponding formula for a Fourier transform will be used below

$$F(\mathbf{q}) = \int f(\mathbf{r}) e^{-i\mathbf{q}\cdot\mathbf{r}} d\mathbf{r}$$

where the integral is summed over all possible values of the position vector **r** within the crystal.

The Fourier transform F(**q**) is generally a complex number, and therefore has a magnitude |F(**q**)| and a phase $\varphi(\mathbf{q})$ related by the equation

$$F(\mathbf{q}) = |F(\mathbf{q})|\, e^{i\phi(\mathbf{q})}$$

The intensities of the reflections observed in X-ray diffraction give us the magnitudes |F(**q**)| but not the phases $\varphi(\mathbf{q})$. To obtain the phases, full sets of reflections are collected with known alterations to the scattering, either by modulating the wavelength past a certain absorption edge or by adding strongly scattering (i.e., electron-dense) metal atoms such as mercury. Combining the magnitudes and phases yields the full Fourier transform F(**q**), which may be inverted to obtain the electron density f(**r**).

Crystals are often idealized as being perfectly periodic. In that ideal case, the atoms are positioned on a perfect lattice, the electron density is perfectly periodic, and the Fourier transform F(**q**) is zero except when **q** belongs to the reciprocal lattice (the so-called Bragg peaks). In reality, however, crystals are not perfectly periodic; atoms vibrate about their mean position, and there may be disorder of various types, such as mosaicity, dislocations, various point defects, and heterogeneity in the conformation of crystallized molecules. Therefore, the Bragg peaks have a finite width and there may be significant diffuse scattering, a continuum of scattered X-rays that fall between the Bragg peaks.

Intuitive understanding by Bragg's law

An intuitive understanding of X-ray diffraction can be obtained from the Bragg model of diffraction. In this model, a given reflection is associated with a set of evenly spaced sheets running through the crystal, usually passing through the centers of the atoms of the crystal lattice. The orientation of a particular set of sheets is identified by its three Miller indices (h, k, l), and let their spacing be noted by d. William Lawrence Bragg proposed a model in which the incoming X-rays are scattered specularly (mirror-like) from each plane; from that assumption, X-rays scattered from adjacent planes will combine constructively (constructive interference) when the angle θ between the plane and the X-ray results in a path-length difference that is an integer multiple n of the X-ray wavelength λ.

$$2d \sin\theta = n\lambda$$

A reflection is said to be indexed when its Miller indices (or, more correctly, its reciprocal lattice vector components) have been identified from the known wavelength and the scattering angle 2θ. Such indexing gives the unit-cell parameters, the lengths and angles of the unit-cell, as well as its space group. Since Bragg's law does not interpret the relative intensities of the reflections, however, it is generally inadequate to solve for the arrangement of atoms within the unit-cell; for that, a Fourier transform method must be carried out.

Scattering as a Fourier Transform

The incoming X-ray beam has a polarization and should be represented as a vector wave; however, for simplicity, let it be represented here as a scalar wave. We also ignore the complication of the time dependence of the wave and just concentrate on the wave's spatial dependence. Plane waves can be represented by a wave vector $\mathbf{k}_{in}$, and so the strength of the incoming wave at time t=0 is given by

$$Ae^{i\mathbf{k}_{in}\cdot\mathbf{r}}$$

At position **r** within the sample, let there be a density of scatterers f(**r**); these scatterers should produce a scattered spherical wave of amplitude proportional to the local amplitude of the incoming wave times the number of scatterers in a small volume dV about **r**

$$\text{amplitude of scattered wave} = Ae^{i\mathbf{k}\cdot\mathbf{r}}Sf(\mathbf{r})dV$$

where S is the proportionality constant.

Let's consider the fraction of scattered waves that leave with an outgoing wave-vector of $\mathbf{k}_{out}$ and strike the screen at $\mathbf{r}_{screen}$. Since no energy is lost (elastic, not inelastic scattering), the wavelengths are the same as are the magnitudes of the wave-vectors $|\mathbf{k}_{in}|=|\mathbf{k}_{out}|$. From the time that the photon is scattered at **r** until it is absorbed at $\mathbf{r}_{screen}$, the photon undergoes a change in phase

$$e^{i\mathbf{k}_{out}\cdot(\mathbf{r}_{screen}-\mathbf{r})}$$

The net radiation arriving at $\mathbf{r}_{screen}$ is the sum of all the scattered waves throughout the crystal

$$AS\int d\mathbf{r}f(\mathbf{r})e^{i\mathbf{k}_{in}\cdot\mathbf{r}}e^{i\mathbf{k}_{out}\cdot(\mathbf{r}_{screen}-\mathbf{r})} = ASe^{i\mathbf{k}_{out}\cdot\mathbf{r}_{screen}}\int d\mathbf{r}f(\mathbf{r})e^{i(\mathbf{k}_{in}-\mathbf{k}_{out})\cdot\mathbf{r}}$$

which may be written as a Fourier transform

$$ASe^{i\mathbf{k}_{out}\cdot\mathbf{r}_{screen}}\int d\mathbf{r}f(\mathbf{r})e^{-i\mathbf{q}\cdot\mathbf{r}} = ASe^{i\mathbf{k}_{out}\cdot\mathbf{r}_{screen}}F(\mathbf{q})$$

where $\mathbf{q} = \mathbf{k}_{out} - \mathbf{k}_{in}$. The measured intensity of the reflection will be square of this amplitude

$$A^2S^2|F(\mathbf{q})|^2$$

Friedel and Bijvoet Mates

For every reflection corresponding to a point **q** in the reciprocal space, there is another reflection of the same intensity at the opposite point -**q**. This opposite reflection is known as the Friedel mate of the original reflection. This symmetry results from the mathematical fact that the density of electrons f(**r**) at a position **r** is always a real number. As noted above, f(**r**) is the inverse transform of its Fourier transform F(**q**); however, such an inverse transform is a complex number in general. To ensure that f(**r**) is real, the Fourier transform F(**q**) must be such that the Friedel mates F(−**q**) and F(**q**) are complex conjugates of one another. Thus, F(−**q**) has the same magnitude as F(**q**) but they have the opposite phase, i.e., φ(**q**) = −φ(**q**)

$$F(-\mathbf{q}) = |F(-\mathbf{q})|e^{i\phi(-\mathbf{q})} = F^*(\mathbf{q}) = |F(\mathbf{q})|e^{-i\phi(\mathbf{q})}$$

The equality of their magnitudes ensures that the Friedel mates have the same intensity $|F|^2$. This symmetry allows one to measure the full Fourier transform from only half the reciprocal space, e.g., by rotating the crystal slightly more than 180° instead of a full 360° revolution. In crystals with significant symmetry, even more reflections may have the same intensity (Bijvoet mates); in such cases, even less of the reciprocal space may need to be measured. In favorable cases of high symmetry, sometimes only 90° or even only 45° of data are required to completely explore the reciprocal space.

$$f(\mathbf{r}) = \int \frac{d\mathbf{q}}{(2\pi)^3} F(\mathbf{q}) e^{i\mathbf{q}\cdot\mathbf{r}} = \int \frac{d\mathbf{q}}{(2\pi)^3} |F(\mathbf{q})| e^{i\phi(\mathbf{q})} e^{i\mathbf{q}\cdot\mathbf{r}}$$

Since Euler's formula states that $e^{ix} = \cos(x) + i\sin(x)$, the inverse Fourier transform can be separated into a sum of a purely real part and a purely imaginary part

$$f(\mathbf{r}) = \int \frac{d\mathbf{q}}{(2\pi)^3} |F(\mathbf{q})| e^{i(\phi+\mathbf{q}\cdot\mathbf{r})} = \int \frac{d\mathbf{q}}{(2\pi)^3} |F(\mathbf{q})| \cos(\phi + \mathbf{q}\cdot\mathbf{r}) + i\int \frac{d\mathbf{q}}{(2\pi)^3} |F(\mathbf{q})| \sin(\phi + \mathbf{q}\cdot\mathbf{r}) = I_{\cos} + iI_{\sin}$$

The function f(**r**) is real if and only if the second integral $I_{\sin}$ is zero for all values of **r**. In turn, this is true if and only if the above constraint is satisfied

$$I_{\sin} = \int \frac{d\mathbf{q}}{(2\pi)^3} |F(\mathbf{q})| \sin(\phi + \mathbf{q}\cdot\mathbf{r}) = \int \frac{d\mathbf{q}}{(2\pi)^3} |F(-\mathbf{q})| \sin(-\phi - \mathbf{q}\cdot\mathbf{r}) = -I_{\sin}$$

since $I_{\sin} = -I_{\sin}$ implies that $I_{\sin}=0$.

Ewald's Sphere

Each X-ray diffraction image represents only a slice, a spherical slice of reciprocal space, as may be seen by the Ewald sphere construction. Both $\mathbf{k}_{out}$ and $\mathbf{k}_{in}$ have the same length, due to the elastic scattering, since the wavelength has not changed. Therefore, they may be represented as two radial vectors in a sphere in reciprocal space, which shows the values of **q** that are sampled in a given diffraction image. Since there is a slight spread in the incoming wavelengths of the incoming X-ray beam, the values of $|F(\mathbf{q})|$ can be measured only for **q** vectors located between the two spheres corresponding to those radii. Therefore, to obtain a full set of Fourier transform data, it is necessary to rotate the crystal through slightly more than 180°, or sometimes less if sufficient symmetry is present. A full 360° rotation is not needed because of a symmetry intrinsic to the Fourier transforms of real functions (such as the electron density), but "slightly more" than 180° is needed to cover all of reciprocal space within a given resolution because of the curvature of the Ewald sphere. In practice, the crystal is rocked by a small amount (0.25-1°) to incorporate reflections near the boundaries of the spherical Ewald's shells.

Patterson Function

A well-known result of Fourier transforms is the autocorrelation theorem, which states that the autocorrelation c(**r**) of a function f(**r**)

$$c(\mathbf{r}) = \int d\mathbf{x} f(\mathbf{x}) f(\mathbf{x}+\mathbf{r}) = \int \frac{d\mathbf{q}}{(2\pi)^3} C(\mathbf{q}) e^{i\mathbf{q}\cdot\mathbf{r}}$$

has a Fourier transform C(**q**) that is the squared magnitude of F(**q**)

$$C(\mathbf{q}) = |F(\mathbf{q})|^2$$

Therefore, the autocorrelation function c(**r**) of the electron density (also known as the Patterson function) can be computed directly from the reflection intensities, without computing the phases. In principle, this could be used to determine the crystal structure directly; however, it is difficult to realize in practice. The autocorrelation function corresponds to the distribution of vectors between atoms in the crystal; thus, a crystal of N atoms in its unit cell may have N(N-1) peaks in its Patterson function. Given the inevitable errors in measuring the intensities, and the mathematical difficulties of reconstructing atomic positions from the interatomic vectors, this technique is rarely used to solve structures, except for the simplest crystals.

Advantages of a Crystal

In principle, an atomic structure could be determined from applying X-ray scattering to non-crystalline samples, even to a single molecule. However, crystals offer a much stronger signal due to their periodicity. A crystalline sample is by definition periodic; a crystal is composed of many unit cells repeated indefinitely in three independent directions. Such periodic systems have a Fourier transform that is concentrated at periodically repeating points in reciprocal space known as Bragg peaks; the Bragg peaks correspond to the reflection spots observed in the diffraction image. Since the amplitude at these reflections grows linearly with the number N of scatterers, the observed intensity of these spots should grow quadratically, like N^2. In other words, using a crystal concentrates the weak scattering of the individual unit cells into a much more powerful, coherent reflection that can be observed above the noise. This is an example of constructive interference.

In a liquid, powder or amorphous sample, molecules within that sample are in random orientations. Such samples have a continuous Fourier spectrum that uniformly spreads its amplitude thereby reducing the measured signal intensity, as is observed in SAXS. More importantly, the orientational information is lost. Although theoretically possible, it is experimentally difficult to obtain atomic-resolution structures of complicated, asymmetric molecules from such rotationally averaged data. An intermediate case is fiber diffraction in which the subunits are arranged periodically in at least one dimension.

NOBEL PRIZES FOR X-RAY CRYSTALLOGRAPHY

Year	Laureate	Prize	Rationale
1914	Max von Laue	Physics	"For his discovery of the diffraction of X-rays by crystals", an important step in the development of X-ray spectroscopy.
1915	William Henry Bragg	Physics	"For their services in the analysis of crystal structure by means of X-rays",
1915	William Lawrence Bragg	Physics	"For their services in the analysis of crystal structure by means of X-rays",
1962	Max F. Perutz	Chemistry	"for their studies of the structures of globular proteins"

1962	John C. Kendrew	Chemistry	"for their studies of the structures of globular proteins"
1962	James Dewey Watson	Medicine	"For their discoveries concerning the molecular structure of nucleic acids and its significance for information transfer in living material"
1962	Francis Harry Compton Crick	Medicine	"For their discoveries concerning the molecular structure of nucleic acids and its significance for information transfer in living material"
1962	Maurice Hugh Frederick Wilkins	Medicine	"For their discoveries concerning the molecular structure of nucleic acids and its significance for information transfer in living material"
1964	Dorothy Hodgkin	Chemistry	"For her determinations by X-ray techniques of the structures of important biochemical substances"
1972	Stanford Moore	Chemistry	"For their contribution to the understanding of the connection between chemical structure and catalytic activity of the active centre of the ribonuclease molecule"
1972	William H. Stein	Chemistry	"For their contribution to the understanding of the connection between chemical structure and catalytic activity of the active centre of the ribonuclease molecule"
1976	William N. Lipscomb	Chemistry	"For his studies on the structure of boranes illuminating problems of chemical bonding"
1985	Jerome Karle	Chemistry	"For their outstanding achievements in developing direct methods for the determination of crystal structures"
1985	Herbert A. Hauptman	Chemistry	"For their outstanding achievements in developing direct methods for the determination of crystal structures"
1988	Johann Deisenhofer	Chemistry	"For their determination of the three-dimensional structure of a photosynthetic reaction centre"
1988	Hartmut Michel	Chemistry	"For their determination of the three-dimensional structure of a photosynthetic reaction centre"
1988	Robert Huber	Chemistry	"For their determination of the three-dimensional structure of a photosynthetic reaction centre"
1997	John E. Walker	Chemistry	"For their elucidation of the enzymatic mechanism underlying the synthesis of adenosine triphosphate (ATP)"
2003	Roderick MacKinnon	Chemistry	"For discoveries concerning channels in cell membranes [...] for structural and mechanistic studies of ion channels"
2003	Peter Agre	Chemistry	"For discoveries concerning channels in cell membranes [...] for the discovery of water channels"
2006	Roger D. Kornberg	Chemistry	"For his studies of the molecular basis of eukaryotic transcription"
2009	Ada E. Yonath	Chemistry	"For studies of the structure and function of the ribosome"
2009	Thomas A. Steitz	Chemistry	"For studies of the structure and function of the ribosome"
2009	Venkatraman Ramakrishnan	Chemistry	"For studies of the structure and function of the ribosome"
2012	Brian Kobilka	Chemistry	"For studies of G-protein-coupled receptors"

MICROSCOPY

Microscopy is the technical field of using microscopes to view objects and areas of objects that cannot be seen with the naked eye (objects that are not within the resolution range of the normal eye). There are three well-known branches of microscopy: optical, electron, and scanning probe microscopy.

Scanning electron microscope image of pollen

Optical & electron microscopy involve the diffraction, reflection, or refraction of electromagnetic radiation/electron beams interacting with the specimen, and the collection of the scattered radiation or another signal in order to create an image. This process may be carried out by wide-field irradiation of the sample (for example standard light microscopy and transmission electron microscopy) or by scanning of a fine beam over the sample (for example confocal laser scanning microscopy and scanning electron microscopy). Scanning probe microscopy involves the interaction of a scanning probe with the surface of the object of interest. The development of microscopy revolutionized biology, gave rise to the field of histology and so remains an essential technique in the life and physical sciences.

Optical Microscopy

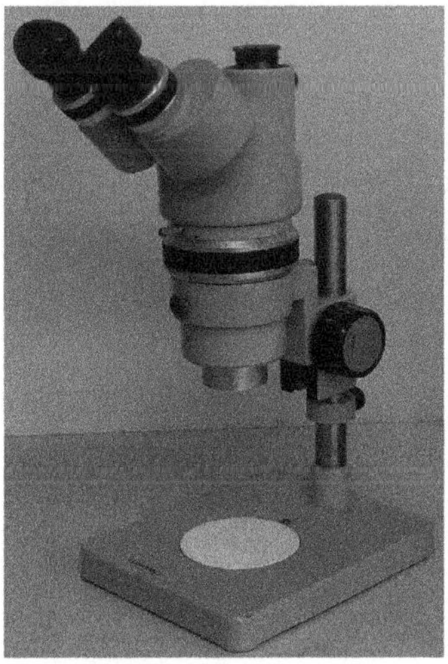

Stereo microscope.

Optical or light microscopy involves passing visible light transmitted through or reflected from the sample through a single or multiple lenses to allow a magnified view of the sample. The resulting image can be detected directly by the eye, imaged on a photographic plate or captured digitally. The single lens with its attachments, or the system of lenses and imaging equipment, along with the appropriate lighting equipment, sample stage and support, makes up the basic light microscope. The most recent development is the digital microscope, which uses a CCD camera to focus on the exhibit of interest. The image is shown on a computer screen, so eye-pieces are unnecessary.

Limitations

Limitations of standard optical microscopy (bright field microscopy) lie in three areas;

- This technique can only image dark or strongly refracting objects effectively.
- Diffraction limits resolution to approximately 0.2 micrometres (see: microscope). This limits the practical magnification limit to ~1500x.
- Out of focus light from points outside the focal plane reduces image clarity.

Live cells in particular generally lack sufficient contrast to be studied successfully, since the internal structures of the cell are colourless and transparent. The most common way to increase contrast is to stain the different structures with selective dyes, but this often involves killing and fixing the sample. Staining may also introduce artifacts, apparent structural details that are caused by the processing of the specimen and are thus not legitimate features of the specimen. In general, these techniques make use of differences in the refractive index of cell structures. It is comparable to looking through a glass window: you (bright field microscopy) don't see the glass but merely the dirt on the glass. There is a difference, as glass is a denser material, and this creates a difference in phase of the light passing through. The human eye is not sensitive to this difference in phase, but clever optical solutions have been thought out to change this difference in phase into a difference in amplitude (light intensity).

Techniques

In order to improve specimen contrast or highlight certain structures in a sample special techniques must be used. A huge selection of microscopy techniques are available to increase contrast or label a sample.

- Four examples of transillumination techniques used to generate contrast in a sample of tissue paper. 1.559 μm/pixel.

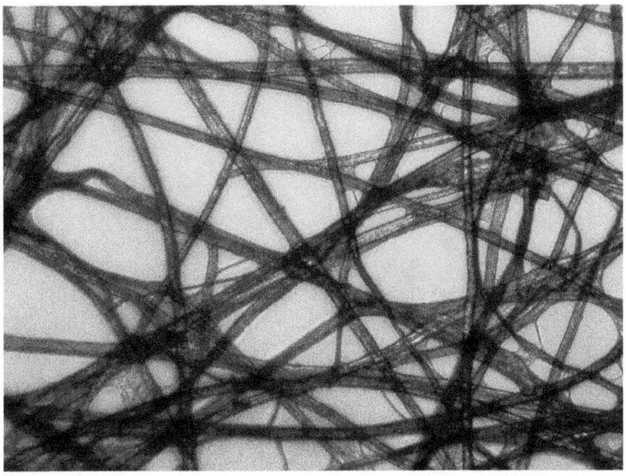

Bright field illumination, sample contrast comes from absorbance of light in the sample.

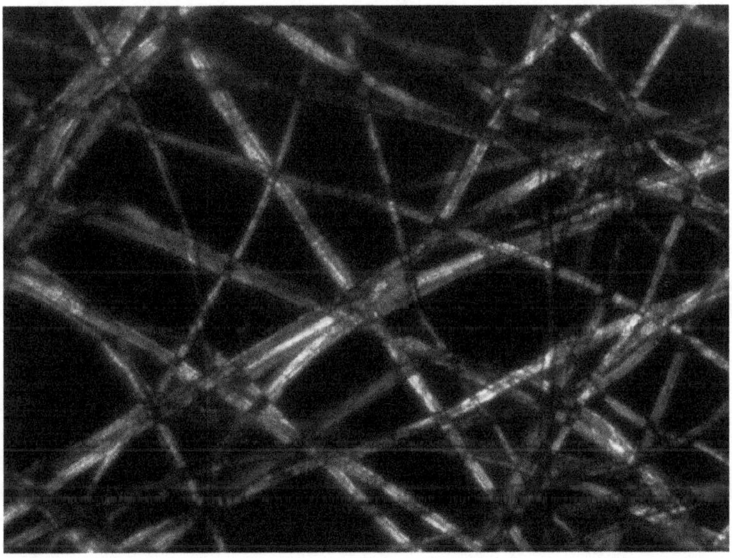

Cross-polarized light illumination, sample contrast comes from rotation of polarized light through the sample.

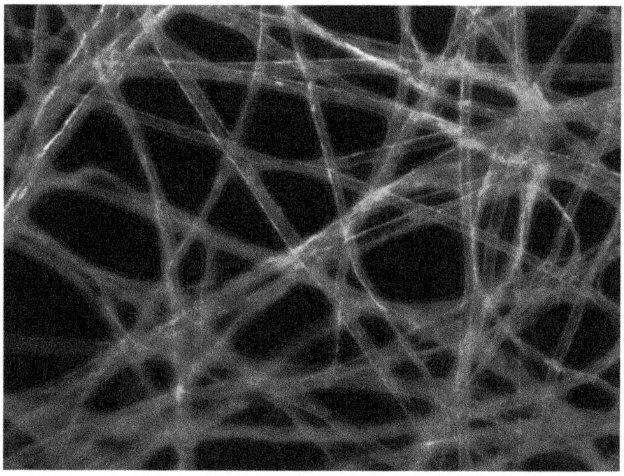

Dark field illumination, sample contrast comes from light scattered by the sample.

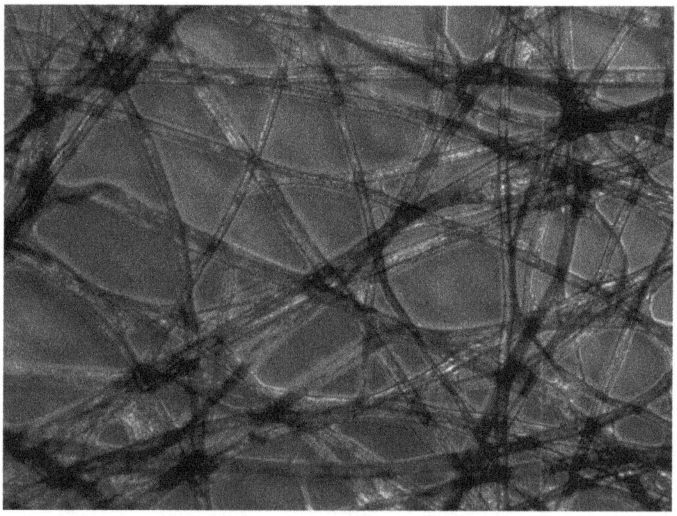

Phase contrast illumination, sample contrast comes from interference of different path lengths of light through the sample.

Bright Field

Bright field microscopy is the simplest of all the light microscopy techniques. Sample illumination is via transmitted white light, i.e. illuminated from below and observed from above. Limitations include low contrast of most biological samples and low apparent resolution due to the blur of out of focus material. The simplicity of the technique and the minimal sample preparation required are significant advantages.

Oblique Illumination

The use of oblique (from the side) illumination gives the image a 3-dimensional appearance and can highlight otherwise invisible features. A more recent technique based on this method is Hoffmann's modulation contrast, a system found on inverted microscopes for use in cell culture. Oblique illumination suffers from the same limitations as bright field microscopy (low contrast of many biological samples; low apparent resolution due to out of focus objects).

Dark field

Dark field microscopy is a technique for improving the contrast of unstained, transparent specimens. Dark field illumination uses a carefully aligned light source to minimize the quantity of directly transmitted (unscattered) light entering the image plane, collecting only the light scattered by the sample. Dark field can dramatically improve image contrast – especially of transparent objects – while requiring little equipment setup or sample preparation. However, the technique suffers from low light intensity in final image of many biological samples, and continues to be affected by low apparent resolution.

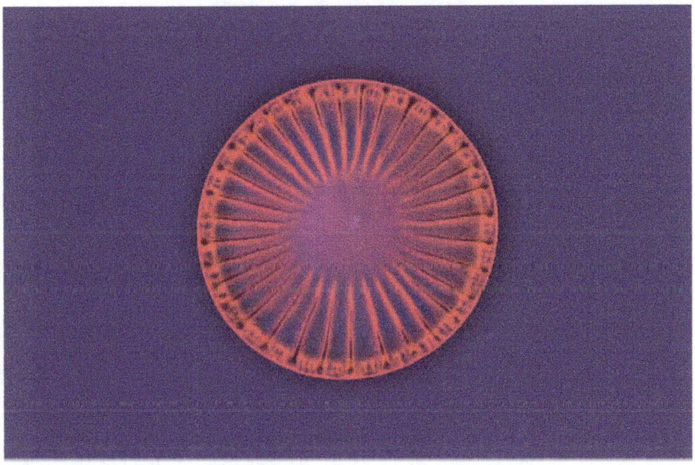

A diatom under Rheinberg illumination

Rheinberg illumination is a special variant of dark field illumination in which transparent, colored filters are inserted just before the condenser so that light rays at high aperture are differently colored than those at low aperture (i.e. the background to the specimen may be blue while the object appears self-luminous red). Other color combinations are possible but their effectiveness is quite variable.

Dispersion Staining

Dispersion staining is an optical technique that results in a colored image of a colorless object. This is an optical staining technique and requires no stains or dyes to produce a color effect. There are five different microscope configurations used in the broader technique of dispersion staining. They include brightfield Becke line, oblique, darkfield, phase contrast, and objective stop dispersion staining.

Phase Contrast

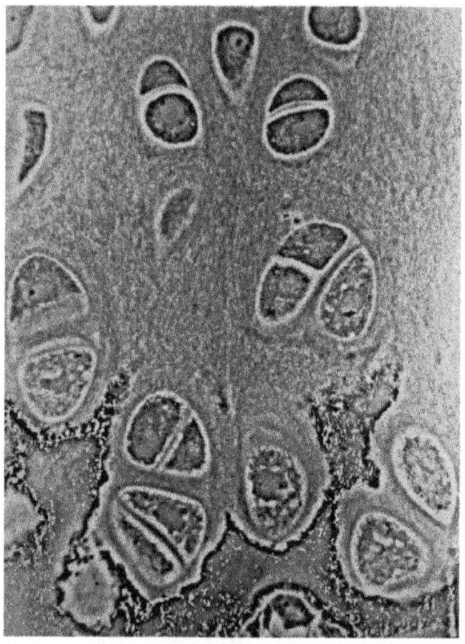

Phase-contrast light micrograph of hyaline cartilage showing chondrocytes and organelles, lacunae and extracellular matrix.

More sophisticated techniques will show proportional differences in optical density. **Phase contrast** is a widely used technique that shows differences in refractive index as difference in contrast. It was developed by the Dutch physicist Frits Zernike in the 1930s (for which he was awarded the Nobel Prize in 1953). The nucleus in a cell for example will show up darkly against the surrounding cytoplasm. Contrast is excellent; however it is not for use with thick objects. Frequently, a halo is formed even around small objects, which obscures detail. The system consists of a circular annulus in the condenser, which produces a cone of light. This cone is superimposed on a similar sized ring within the phase-objective. Every objective has a different size ring, so for every objective another condenser setting has to be chosen. The ring in the objective has special optical properties: it, first of all, reduces the direct light in intensity, but more importantly, it creates an artificial phase difference of about a quarter wavelength. As the physical properties of this direct light have changed, interference with the diffracted light occurs, resulting in the phase contrast image. One disadvantage of phase-contrast microscopy is halo formation (halo-light ring).

Differential interference contrast

Superior and much more expensive is the use of **interference contrast**. Differences in optical density will show up as differences in relief. A nucleus within a cell will actually show up as a globule in the most often used **differential interference contrast** system according to Georges Nomarski. However, it has to be kept in mind that this is an optical effect, and the relief does not necessarily resemble the true shape. Contrast is very good and the condenser aperture can be used fully open, thereby reducing the depth of field and maximizing resolution.

The system consists of a special prism (Nomarski prism, Wollaston prism) in the condenser that splits light in an ordinary and an extraordinary beam. The spatial difference between the two beams is minimal (less than the maximum resolution of the objective). After passage through the specimen, the beams are reunited by a similar prism in the objective.

In a homogeneous specimen, there is no difference between the two beams, and no contrast is being generated. However, near a refractive boundary (say a nucleus within the cytoplasm), the difference between the ordinary and the extraordinary beam will generate a relief in the image. Differential interference contrast requires a polarized light source to function; two polarizing filters have to be fitted in the light path, one below the condenser (the polarizer), and the other above the objective (the analyzer).

Note: In cases where the optical design of a microscope produces an appreciable lateral separation of the two beams we have the case of classical interference microscopy, which does not result in relief images, but can nevertheless be used for the quantitative determination of mass-thicknesses of microscopic objects.

Interference Reflection

An additional technique using interference is **interference reflection microscopy** (also known as reflected interference contrast, or RIC). It relies on cell adhesion to the slide to produce an interference signal. If there is no cell attached to the glass, there will be no interference.

Interference reflection microscopy can be obtained by using the same elements used by DIC, but without the prisms. Also, the light that is being detected is reflected and not transmitted as it is when DIC is employed.

Fluorescence

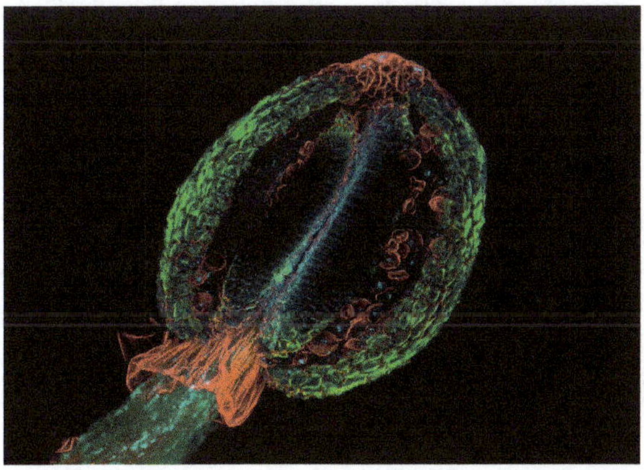

Images may also contain artifacts. This is a confocal laser scanning fluorescencemicrograph of thale cress anther (part of stamen). The picture shows among other things a nice red flowing collar-like structure just below the anther. However, an intact thale cress stamen does not have such collar, this is a fixation artifact: the stamen has been cut below the picture frame, and epidermis (upper layer of cells) of stamen stalk has peeled off, forming a non-characteristic structure. Photo: Heiti Paves from Tallinn University of Technology.

When certain compounds are illuminated with high energy light, they emit light of a lower frequency. This effect is known as fluorescence. Often specimens show their characteristic autofluorescence image, based on their chemical makeup.

This method is of critical importance in the modern life sciences, as it can be extremely sensitive, allowing the detection of single molecules. Many different fluorescent dyes can be used to stain different structures or chemical compounds. One particularly powerful method is the combination of antibodies coupled to a fluorophore as in immunostaining. Examples of commonly used fluorophores are fluorescein or rhodamine.

The antibodies can be tailor-made for a chemical compound. For example, one strategy often in use is the artificial production of proteins, based on the genetic code (DNA). These proteins can then be used to immunize rabbits, forming antibodies which bind to the protein. The antibodies are then coupled chemically to a fluorophore and used to trace the proteins in the cells under study.

Highly efficient fluorescent proteins such as the green fluorescent protein (GFP) have been developed using the molecular biology technique of gene fusion, a process that links the expression of the fluorescent compound to that of the target protein. This combined fluorescent protein is, in general, non-toxic to the organism and rarely interferes with the function of the protein under study. Genetically modified cells or organisms directly express the fluorescently tagged proteins, which enables the study of the function of the original protein in vivo.

Growth of protein crystals results in both protein and salt crystals. Both are colorless and microscopic. Recovery of the protein crystals requires imaging which can be done by the intrinsic fluorescence of the protein or by using transmission microscopy. Both methods require an ultraviolet microscope as protein absorbs light at 280 nm. Protein will also fluorescence at approximately 353 nm when excited with 280 nm light.

Since fluorescence emission differs in wavelength (color) from the excitation light, an ideal fluorescent image shows only the structure of interest that was labeled with the fluorescent dye. This high specificity led to the widespread use of fluorescence light microscopy in biomedical research. Different fluorescent dyes can be used to stain different biological structures, which can then be detected simultaneously, while still being specific due to the individual color of the dye.

To block the excitation light from reaching the observer or the detector, filter sets of high quality are needed. These typically consist of an excitation filter selecting the range of excitation wavelengths, a dichroic mirror, and an emission filter blocking the excitation light. Most fluorescence microscopes are operated in the Epi-illumination mode (illumination and detection from one side of the sample) to further decrease the amount of excitation light entering the detector.

An example of fluorescence microscopy today is two-photon or multi-photon imaging. Two photon imaging allows imaging of living tissues up to a very high depth by enabling greater excitation light penetration and reduced background emission signal.

Confocal

Confocal microscopy uses a scanning point of light and a pinhole to prevent out of focus light from reaching the detector. Compared to full sample illumination, confocal microscopy gives slightly higher resolution, and significantly improves optical sectioning. Confocal microscopy is, therefore, commonly used where 3D structure is important.

Single Plane Illumination microscopy and light Sheet Fluorescence Microscopy

Using a plane of light formed by focusing light through a cylindrical lens at a narrow angle or by scanning a line of light in a plane perpendicular to the axis of objective, high resolution optical sections can be taken. Single plane illumination, or light sheet illumination, is also accomplished using beam shaping techniques incorporating multiple-prism beam expanders. The images are captured by CCDs. These variants allow very fast and high signal to noise ratio image capture.

Wide-Field Multiphoton Microscopy

Wide-field multiphoton microscopy refers to an optical non-linear imaging technique tailored for ultrafast imaging in which a large area of the object is illuminated and imaged without the need for scanning. High intensities are required to induce non-linear optical processes such as two-photon fluorescence or second harmonic generation. In scanning multiphoton microscopes the high intensities are achieved by tightly focusing the light, and the image is obtained by stage- or beam-scanning the sample. In **wide-field multiphoton microscopy** the high intensities are best achieved using an optically amplified pulsed laser source to attain a large field of view (~100 μm). The image in this case is obtained as a single frame with a CCD without the need of scanning, making the technique particularly useful to visualize dynamic processes simultaneously across the object of interest. With wide-field multiphoton microscopy the frame rate can be increased up to a 1000-fold compared to multiphoton scanning microscopy.

Deconvolution

Fluorescence microscopy is a powerful technique to show specifically labeled structures within a complex environment and to provide three-dimensional information of biological structures. However, this information is blurred by the fact that, upon illumination, all fluorescently labeled structures emit light, irrespective of whether they are in focus or not. So an image of a certain structure is always blurred by the contribution of light from structures that are out of focus. This phenomenon results in a loss of contrast especially when using objectives with a high resolving power, typically oil immersion objectives with a high numerical aperture.

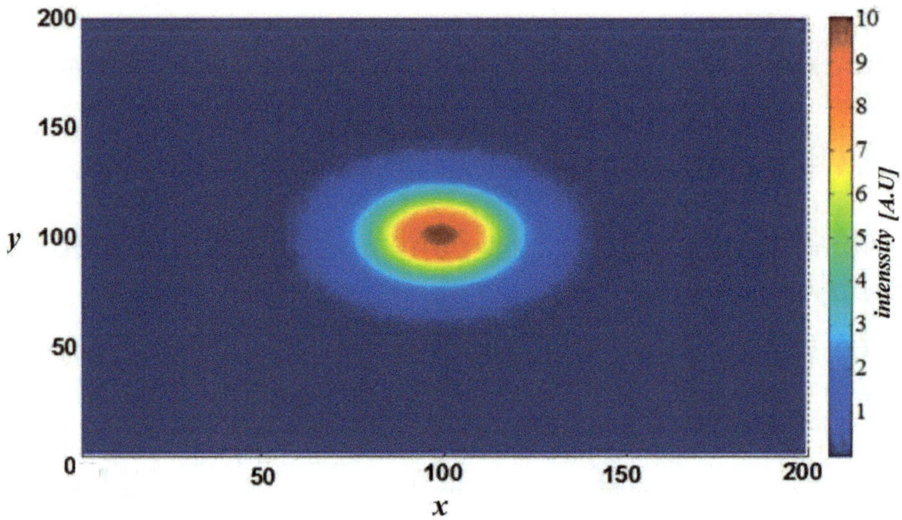

Mathematically modeled Point Spread Function of a pulsed THz laser imaging system.

However, blurring is not caused by random processes, such as light scattering, but can be well defined by the optical properties of the image formation in the microscope imaging system. If one considers a small fluorescent light source (essentially a bright spot), light coming from this spot spreads out further from our perspective as the spot becomes more out of focus. Under ideal conditions, this produces an "hourglass" shape of this point source in the third (axial) dimension. This shape is called the point spread function (PSF) of the microscope imaging system. Since any fluorescence image is made up of a large number of such small fluorescent light sources, the image is said to be "convolved by the point spread function". The mathematically modeled PSF of a terahertz laser pulsed imaging system is shown on the right.

The output of an imaging system can be described using the equation:

$$s(x, y) = PSF(x, y) * o(x, y) + n$$

Where n is the additive noise. Knowing this point spread function means that it is possible to reverse this process to a certain extent by computer-based methods commonly known as deconvolution microscopy. There are various algorithms available for 2D or 3D deconvolution. They can be roughly classified in nonrestorative and restorative methods. While the nonrestorative methods can improve contrast by removing out-of-focus light from focal planes, only the restorative methods can actually reassign light to its proper place of origin. Processing fluorescent images in this manner can be an advantage over directly acquiring images without out-of-focus light, such as images from confocal microscopy, because light signals otherwise eliminated become useful information. For 3D deconvolution, one typically provides a series of images taken from different focal planes (called a Z-stack) plus the knowledge of the PSF, which can be derived either experimentally or theoretically from knowing all contributing parameters of the microscope.

Sub-Diffraction Techniques

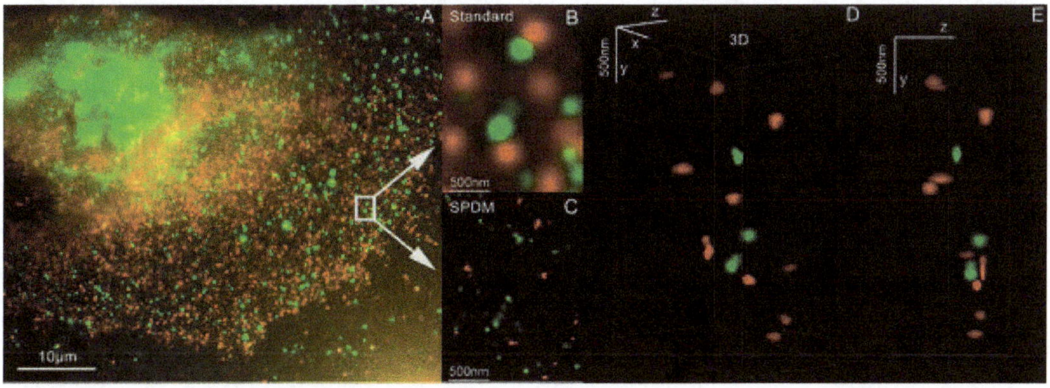

Example of super-resolution microscopy. Image of Her3 and Her2, target of the breast cancer drugTrastuzumab, within a cancer cell.

A multitude of super-resolution microscopy techniques have been developed in recent times which circumvent the diffraction barrier.

This is mostly achieved by imaging a sufficiently static sample multiple times and either modifying the excitation light or observing stochastic changes in the image.

Knowledge of and chemical control over fluorophore photophysics is at the core of these techniques, by which resolutions of ~20 nanometers are regularly obtained.

Serial time-Encoded Amplified Microscopy

Serial time encoded amplified microscopy (STEAM) is an imaging method that provides ultrafast shutter speed and frame rate, by using optical image amplification to circumvent the fundamental trade-off between sensitivity and speed, and a single-pixel photodetector to eliminate the need for a detector array and readout time limitations The method is at least 1000 times faster than the state-of-the-art CCD and CMOS cameras. Consequently, it is potentially useful for a broad range of scientific, industrial, and biomedical applications that require high image acquisition rates, including real-time diagnosis and evaluation of shockwaves, microfluidics, MEMS, and laser surgery.

Extensions

Most modern instruments provide simple solutions for micro-photography and image recording electronically. However such capabilities are not always present and the more experienced microscopist will, in many cases, still prefer a hand drawn image to a photograph. This is because a microscopist with knowledge of the subject can accurately convert a three-dimensional image into a precise two-dimensional drawing. In a photograph or other image capture system however, only one thin plane is ever in good focus.

The creation of careful and accurate micrographs requires a microscopical technique using a monocular eyepiece. It is essential that both eyes are open and that the eye that is not observing down the microscope is instead concentrated on a sheet of paper on the bench besides the microscope. With practice, and without moving the head or eyes, it is possible to accurately record the observed details by tracing round the observed shapes by simultaneously "seeing" the pencil point in the microscopical image.

Practicing this technique also establishes good general microscopical technique. It is always less tiring to observe with the microscope focused so that the image is seen at infinity and with both eyes open at all times.

Other Enhancements

Microspectroscopy:spectroscopy with a microscope

X-Ray

As resolution depends on the wavelength of the light. Electron microscopy has been developed since the 1930s that use electron beams instead of light. Because of the much smaller wavelength of the electron beam, resolution is far higher.

Though less common, X-ray microscopy has also been developed since the late 1940s. The resolution of X-ray microscopy lies between that of light microscopy and electron microscopy.

Electron Microscopy

Until the invention of sub-diffraction microscopy, the wavelength of the light limited the resolution of traditional microscopy to around 0.2 micrometers. In order to gain higher resolution, the use of an electron beam with a far smaller wavelength is used in electron microscopes.

- Transmission electron microscopy (TEM) is quite similar to the compound light microscope, by sending an electron beam through a very thin slice of the specimen. The resolution limit in 2005 was around 0.05 nanometer and has not increased appreciably since that time.
- Scanning electron microscopy (SEM) visualizes details on the surfaces of specimens and gives a very nice 3D view. It gives results much like those of the stereo light microscope. The best resolution for SEM in 2011 was 0.4 nanometer.

Electron microscopes equipped for X-ray spectroscopy can provide qualitative and quantitative elemental analysis.

Scanning Probe Microscopy

This is a sub-diffraction technique. Examples of scanning probe microscopes are the atomic force microscope (AFM), the Scanning tunneling microscope, the photonic force microscope and the recurrence tracking microscope. All such methods use the physical contact of a solid probe tip to scan the surface of an object, which is supposed to be almost flat.

Ultrasonic Force

Ultrasonic Force Microscopy (UFM) has been developed in order to improve the details and image contrast on "flat" areas of interest where AFM images are limited in contrast. The combination of AFM-UFM allows a near field acoustic microscopic image to be generated. The AFM tip is used to detect the ultrasonic waves and overcomes the limitation of wavelength that occurs in acoustic microscopy. By using the elastic changes under the AFM tip, an image of much greater detail than the AFM topography can be generated.

Ultrasonic force microscopy allows the local mapping of elasticity in atomic force microscopy by the application of ultrasonic vibration to the cantilever or sample. In an attempt to analyze the results of ultrasonic force microscopy in a quantitative fashion, a force-distance curve measurement is done with ultrasonic vibration applied to the cantilever base, and the results are compared with a model of the cantilever dynamics and tip-sample interaction based on the finite-difference technique.

Ultraviolet Microscopy

Ultraviolet microscopes have two main purposes. The first is to utilize the shorter wavelength of ultraviolet electromagnetic energy to improve the image resolution beyond that of the diffraction limit of standard optical microscopes. This technique is used for non-destructive inspection of devices with very small features such as those found in modern semiconductors. The second application for UV microscopes is contrast enhancement where the response of individual samples is enhanced, relative to their surrounding, due to the interaction of light with the molecules within the sample itself. One example is in the growth of protein crystals. Protein crystals are formed in salt solutions. As salt and protein crystals are both formed in the growth process, and both are commonly transparent to the human eye, they cannot be differentiated with a standard optical microscope. As the tryptophan of protein absorbs light at 280 nm, imaging with a UV microscope with 280 nm bandpass filters makes it simple to differentiate between the two types of crystals. The protein crystals appear dark while the salt crystals are transparent.

Infrared Microscopy

The term infrared microscopy refers to microscopy performed at infrared wavelengths. In the typical instrument configuration a Fourier Transform Infrared Spectrometer (FTIR) is combined with an optical microscope and an infrared detector. The infrared detector can be a single point detector, a linear array or a 2D focal plane array. The FTIR provides the ability to perform chemical analysis via infrared spectroscopy and the microscope and point or array detector enable this chemical analysis to be spatially resolved, i.e. performed at different regions of the sample. As such, the technique is also called infrared microspectroscopy (an alternative architecture involves the combination of a tuneable infrared light source and single point detector on a flying objective). This technique is frequently used for infrared chemical imaging, where the image contrast is determined by the response of individual sample regions to particular IR wavelengths selected by the user, usually specific IR absorption bands and associated molecular resonances . A key limitation of conventional infrared microspectroscopy is that the spatial resolution is diffraction-limited. Specifically the spatial resolution is limited to a figure related to the wavelength of the light. For practical IR microscopes, the spatial resolution is limited to 1-3X the wavelength, depending on the specific technique and instrument used. For mid-IR wavelengths, this sets a practical spatial resolution limit of ~3-30 μm.

IR versions of sub-diffraction microscopy (see above) also exist . These include IR NSOM, photothermal microspectroscopy, and atomic force microscope based infrared spectroscopy (AFM-IR).

Digital Holographic Microscopy

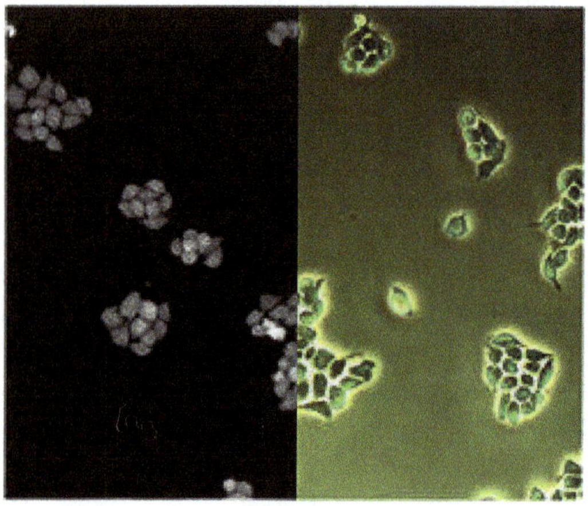

Human cells imaged by DHM phase shift (left) and phase contrast microscopy (right).

In digital holographic microscopy (DHM), interfering wave fronts from a coherent (monochromatic) light-source are recorded on a sensor. The image is digitally reconstructed by a computer from the recorded hologram. Besides the ordinary bright field image, a phase shift image is created.

DHM can operate both in reflection and transmission mode. In reflection mode, the phase shift image provides a relative distance measurement and thus represents a topography map of the reflecting surface. In transmission mode, the phase shift image provides a label-free quantitative measurement of the optical thickness of the specimen. Phase shift images of biological cells are very similar to images of stained cells and have successfully been analyzed by high content analysis software.

A unique feature of DHM is the ability to adjust focus after the image is recorded, since all focus planes are recorded simultaneously by the hologram. This feature makes it possible to image moving particles in a volume or to rapidly scan a surface. Another attractive feature is DHM's ability to use low cost optics by correcting optical aberrations by software.

Digital Pathology (Virtual Microscopy)

Digital pathology is an image-based information environment enabled by computer technology that allows for the management of information generated from a digital slide. Digital pathology is enabled in part by virtual microscopy, which is the practice of converting glass slides into digital slides that can be viewed, managed, and analyzed.

Laser microscopy

Laser microscopy is a rapidly growing field that uses laser illumination sources in various forms of microscopy. For instance, laser microscopy focused on biological applications uses ultrashort pulse lasers, in a number of techniques labeled as nonlinear microscopy, saturation microscopy, and two-photon excitation microscopy.

High-intensity, short-pulse laboratory x-ray lasers have been under development for several years. When this technology comes to fruition, it will be possible to obtain magnified three-dimensional images of elementary biological structures in the living state at a precisely defined instant. For optimum contrast between water and protein and for best sensitivity and resolution, the laser should be tuned near the nitrogen line at about 0.3 nanometers. Resolution will be limited mainly by the hydrodynamic expansion that occurs while the necessary number of photons is being registered. Thus, while the specimen is destroyed by the exposure, its configuration can be captured before it explodes.

Scientists have been working on practical designs and prototypes for x-ray holographic microscopes, despite the prolonged development of the appropriate laser.

Amateur Microscopy

Amateur Microscopy is the investigation and observation of biological and non-biological specimens for recreational purposes. Collectors of minerals, insects, seashells, and plants may use microscopes as tools to uncover features that help them classify their collected items. Other amateurs may be interested in observing the life found in pond water and of other samples. Microscopes may also prove useful for the water quality assessment for people that keep a home aquarium. Photographic documentation and drawing of the microscopic images are additional tasks that augment the spectrum of tasks of the amateur. There are even competitions for photomicrograph art. Participants of this pastime may either use commercially prepared microscopic slides or engage in the task of specimen preparation.

While microscopy is a central tool in the documentation of biological specimens, it is, in general, insufficient to justify the description of a new species based on microscopic investigations alone. Often genetic and biochemical tests are necessary to confirm the discovery of a new species. A laboratory and access to academic literature is a necessity, which is specialized and, in general, not available to amateurs. There is, however, one huge advantage that amateurs have above professionals: time to explore their surroundings. Often, advanced amateurs team up with professionals to validate their findings and (possibly) describe new species.

In the late 1800s, amateur microscopy became a popular hobby in the United States and Europe. Several ‹professional amateurs’ were being paid for their sampling trips and microscopic explorations by philanthropists, to keep them amused on the Sunday afternoon (e.g., the diatom specialist A. Grunow, being paid by (among others) a Belgian industrialist). Professor John Phin published “Practical Hints on the Selection and Use of the Microscope (Second Edition, 1878),” and was also the editor of the “American Journal of Microscopy.”

POWDER DIFFRACTION

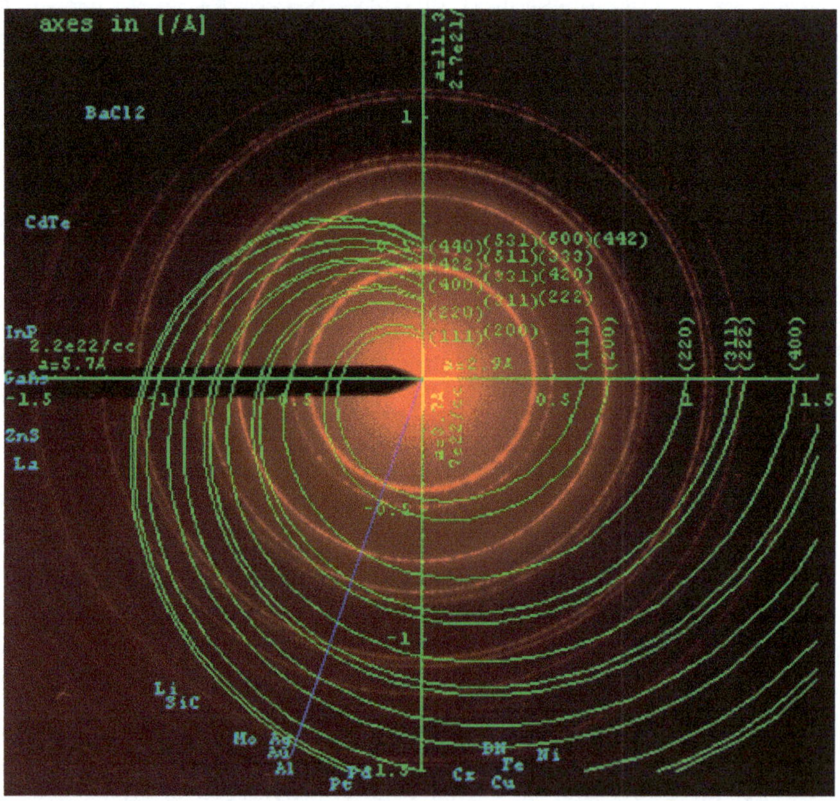

Electron powder pattern (red) of an Al film with an fcc spiral overlay (green) and a line of intersections (blue) that determines lattice parameter.

Powder diffraction is a scientific technique using X-ray, neutron, or electron diffraction on powder or microcrystalline samples for structural characterization of materials. An instrument dedicated to performing such powder measurements is called a **powder diffractometer**.

Powder diffraction stands in contrast to single crystal diffraction techniques, which work best with a single, well-ordered crystal.

Explanation

A diffractometer produces waves at a known frequency, which is determined by their source. The source is often x-rays, because they are the only kind of light with the correct frequency for inter-atomic-scale diffraction. However, electrons and neutrons are also common sources, with their frequency determined by their de Broglie wavelength. When these waves reach the sample, the atoms of the sample act just like a diffraction grating, producing bright spots at particular angles. By measuring the angle where these bright spots occur, the spacing of the diffraction grating can be determined by Bragg's law. Because the sample itself is the diffraction grating, this spacing is the atomic spacing.

The distinction between powder and single crystal diffraction is the degree of texturing in the sample. Single crystals have maximal texturing, and are said to be anisotropic. In contrast, in powder diffraction, every possible crystalline orientation is represented equally in a powdered sample, the isotropic case. PXRD operates under the assumption that the sample is randomly arranged. Therefore, a statistically significant number of each plane of the crystal structure will be in the proper orientation to diffract the X-rays. Therefore, each plane will be represented in the signal. In practice, it is sometimes necessary to rotate the sample orientation to eliminate the effects of texturing and achieve true randomness.

Mathematically, crystals can be described by a Bravais lattice with some regularity in the spacing between atoms. Because of this regularity, we can describe this structure in a different way using the reciprocal lattice, which is related to the original structure by a Fourier transform. This three-dimensional space can be described with reciprocal axes x^*, y^*, and z^* or alternatively in spherical coordinates q, φ^*, and χ^*. In powder diffraction, intensity is homogeneous over φ^* and χ^*, and only q remains as an important measurable quantity. This is because orientational averaging causes the three-dimensional reciprocal space that is studied in single crystal diffraction to be projected onto a single dimension.

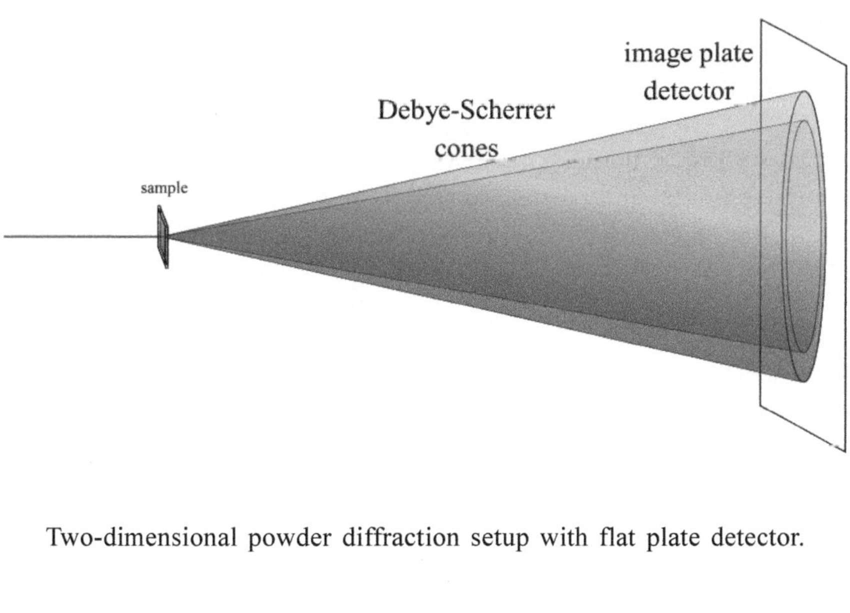

Two-dimensional powder diffraction setup with flat plate detector.

When the scattered radiation is collected on a flat plate detector, the rotational averaging leads to smooth diffraction rings around the beam axis, rather than the discrete Laue spots observed in single crystal diffraction. The angle between the beam axis and the ring is called the scattering angle and in X-ray crystallography always denoted as 2θ (in scattering of visible light the convention is usually to call it θ). In accordance with Bragg's law, each ring corresponds to a particular reciprocal lattice vector G in the sample crystal. This leads to the definition of the scattering vector as:

$$|G| = q = 2k\sin(\theta) = \frac{4\pi}{\lambda}\sin(\theta).$$

In this equation, G is the reciprocal lattice vector, q is the length of the reciprocal lattice vector, k is the momentum transfer vector, θ is half of the scattering angle, and λ is the wavelength of the radiation. Powder diffraction data are usually presented as a diffractogram in which the diffracted intensity, I, is shown as a function either of the scattering angle 2θ or as a function of the scattering vector length q. The latter variable has the advantage that the diffractogram no longer depends on the value of the wavelength λ. The advent of synchrotron sources has widened the choice of wavelength considerably. To facilitate comparability of data obtained with different wavelengths the use of q is therefore recommended and gaining acceptability.

Uses

Relative to other methods of analysis, powder diffraction allows for rapid, non-destructive analysis of multi-component mixtures without the need for extensive sample preparation. This gives laboratories around the world the ability to quickly analyze unknown materials and perform materials characterization in such fields as metallurgy, mineralogy, forensic science, archeology, condensed matter physics, and the biological and pharmaceutical sciences. Identification is performed by comparison of the diffraction pattern to a known standard or to a database such as the International Centre for Diffraction Data's Powder Diffraction File (PDF) or the Cambridge Structural Database (CSD). Advances in hardware and software, particularly improved optics and fast detectors, have dramatically improved the analytical capability of the technique, especially relative to the speed of the analysis. The fundamental physics upon which the technique is based provides high precision and accuracy in the measurement of interplanar spacings, sometimes to fractions of an Ångström, resulting in authoritative identification frequently used in patents, criminal cases and other areas of law enforcement. The ability to analyze multiphase materials also allows analysis of how materials interact in a particular matrix such as a pharmaceutical tablet, a circuit board, a mechanical weld, a geologic core sampling, cement and concrete, or a pigment found in an historic painting. The method has been historically used for the identification and classification of minerals, but it can be used for any materials, even amorphous ones, so long as a suitable reference pattern is known or can be constructed.

Phase Identification

The most widespread use of powder diffraction is in the identification and characterization of crystalline solids, each of which produces a distinctive diffraction pattern. Both the positions (corresponding to lattice spacings) and the relative intensity of the lines in a diffraction pattern are indicative of a particular phase and material, providing a "fingerprint" for comparison. A multi-phase mixture, e.g. a soil sample, will show more than one pattern superposed, allowing for determination of the relative concentrations of phases in the mixture.

J.D. Hanawalt, an analytical chemist who worked for Dow Chemical in the 1930s, was the first to realize the analytical potential of creating a database. Today it is represented by the Powder Diffraction File (PDF) of the International Centre for Diffraction Data (formerly Joint Committee for Powder Diffraction Studies). This has been made searchable by computer through the work of global software developers and equipment manufacturers. There are now over 550,000 reference materials in the 2006 Powder Diffraction File Databases, and these databases are interfaced to a wide variety of diffraction analysis software and distributed globally. The Powder Diffraction File contains many subfiles, such as minerals, metals and alloys, pharmaceuticals, forensics, excipients, superconductors, semiconductors, etc., with large collections of organic, organometallic and inorganic reference materials.

Crystallinity

In contrast to a crystalline pattern consisting of a series of sharp peaks, amorphous materials (liquids, glasses etc.) produce a broad background signal. Many polymers show semicrystalline behavior, i.e. part of the material forms an ordered crystallite by folding of the molecule. A single polymer molecule may well be folded into two different, adjacent crystallites and thus form a tie between the two. The tie part is prevented from crystallizing. The result is that the crystallinity will never reach 100%. Powder XRD can be used to determine the crystallinity by comparing the integrated intensity of the background pattern to that of the sharp peaks. Values obtained from powder XRD are typically comparable but not quite identical to those obtained from other methods such as DSC.

Lattice parameters

The position of a diffraction peak is 'independent' of the atomic positions within the cell and entirely determined by the size and shape of the unit cell of the crystalline phase. Each peak represents a certain lattice plane and can therefore be characterized by a Miller index. If the symmetry is high, e.g.: cubic or hexagonal it is usually not too hard to identify the index of each peak, even for an unknown phase. This is particularly important in solid-state chemistry, where one is interested in finding and identifying new materials. Once a pattern has been indexed, this characterizes the reaction product and identifies it as a new solid phase. Indexing programs exist to deal with the harder cases, but if the unit cell is very large and the symmetry low (triclinic) success is not always guaranteed.

Expansion Tensors, Bulk Modulus

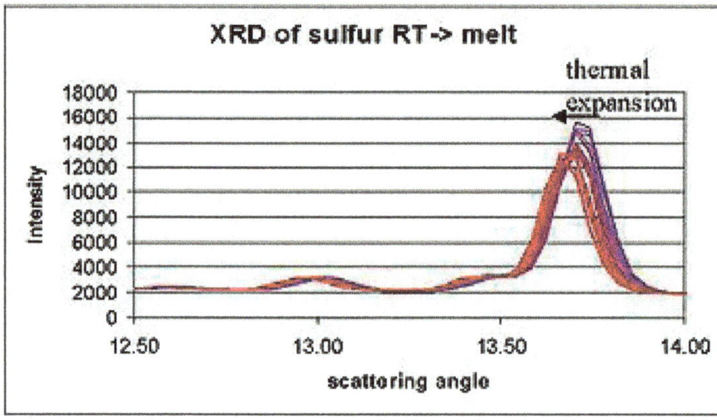

Thermal expansion of a sulfur powder.

Cell parameters are somewhat temperature and pressure dependent. Powder diffraction can be combined with in situ temperature and pressure control. As these thermodynamic variables are changed, the observed diffraction peaks will migrate continuously to indicate higher or lower lattice spacings as the unit cell distorts. This allows for measurement of such quantities as the thermal expansion tensor and the isothermal bulk modulus, as well determination of the full equation of state of the material.

Phase Transitions

At some critical set of conditions, for example 0 °C for water at 1 atm, a new arrangement of atoms or molecules may become stable, leading to a phase transition. At this point new diffraction peaks will appear or old ones disappear according to the symmetry of the new phase. If the material melts to an isotropic liquid, all sharp lines will disappear and be replaced by a broad amorphous pattern. If the transition produces another crystalline phase, one set of lines will suddenly be replaced by another set. In some cases however lines will split or coalesce, e.g. if the material undergoes a continuous, second order phase transition. In such cases the symmetry may change because the existing structure is distorted rather than replaced by a completely different one. For example, the diffraction peaks for the lattice planes (100) and (001) can be found at two different values of q for a tetragonal phase, but if the symmetry becomes cubic the two peaks will come to coincide.

Crystal Structure Refinement and Determination

Crystal structure determination from powder diffraction data is extremely challenging due to the overlap of reflections in a powder experiment. A number of different methods exist for structural determination, such as simulated annealing and charge flipping. The crystal structures of known materials can be refined, i.e. as a function of temperature or pressure, using the Rietveld method. The Rietveld method is a so-called full pattern analysis technique. A crystal structure, together with instrumental and microstructural information, is used to generate a theoretical diffraction pattern that can be compared to the observed data. A least squares procedure is then used to minimize the difference between the calculated pattern and each point of the observed pattern by adjusting model parameters. Techniques to determine unknown structures from powder data do exist, but are somewhat specialized. A number of programs that can be used in structure determination are TOPAS, Fox, DASH, GSAS, EXPO2004, and a few others.

Size and Strain Broadening

There are many factors that determine the width B of a diffraction peak. These include:

- instrumental factors
- the presence of defects to the perfect lattice
- differences in strain in different grains
- the size of the crystallites

It is often possible to separate the effects of size and strain. Where size broadening is independent of q (K=1/d), strain broadening increases with increasing q-values. In most cases there will be both size and strain broadening. It is possible to separate these by combining the two equations in what is known as the Hall–Williamson method:

$$B \cdot \cos(\theta) = \frac{k\lambda}{D} + \eta \cdot \sin(\theta),$$

Thus, when we plot $B \cdot \cos(\theta)$ vs. $\sin(\theta)$ we get a straight line with slope η and intercept $\frac{k\lambda}{D}$. The expression is a combination of the Scherrer equation for size broadening and the Stokes and Wilson expression for strain broadening. The value of η is the strain in the crystallites, the value of D represents the size of the crystallites. The constant k is typically close to unity and ranges from 0.8 to 1.39.

Comparison of X-ray and Neutron Scattering

X-ray photons scatter by interaction with the electron cloud of the material, neutrons are scattered by the nuclei. This means that, in the presence of heavy atoms with many electrons, it may be difficult to detect light atoms by X-ray diffraction. In contrast, the neutron scattering lengths of most atoms are approximately equal in magnitude. Neutron diffraction techniques may therefore be used to detect light elements such as oxygen or hydrogen in combination with heavy atoms. The neutron diffraction technique therefore has obvious applications to problems such as determining oxygen displacements in materials like high temperature superconductors and ferroelectrics, or to hydrogen bonding in biological systems.

A further complication in the case of neutron scattering from hydrogenous materials is the strong incoherent scattering of hydrogen (80.27(6) barn). This leads to a very high background in neutron diffraction experiments, and may make structural investigations impossible. A common solution is deuteration, i.e., replacing the 1-H atoms in the sample with deuterium (2-H). The incoherent scattering length of deuterium is much smaller (2.05(3) barn) making structural investigations significantly easier. However, in some systems, replacing hydrogen with deuterium may alter the structural and dynamic properties of interest.

As neutrons also have a magnetic moment, they are additionally scattered by any magnetic moments in a sample. In the case of long range magnetic order, this leads to the appearance of new Bragg reflections. In most simple cases, powder diffraction may be used to determine the size of the moments and their spatial orientation.

Aperiodically Arranged Clusters

Predicting the scattered intensity in powder diffraction patterns from gases, liquids, and randomly distributed nano-clusters in the solid state is (to first order) done rather elegantly with the Debye scattering equation:

$$I(q) = \sum_{i=1}^{N} \sum_{j=1}^{N} f_i(q) f_j(q) \frac{\sin(q r_{ij})}{q r_{ij}},$$

where the magnitude of the scattering vector q is in reciprocal lattice distance units, N is the number of atoms, $f_i(q)$ is the atomic scattering factor for atom i and scattering vector q, while r_{ij} is the distance between atom i and atom j. One can also use this to predict the effect of nano-crystallite shape on detected diffraction peaks, even if in some directions the cluster is only one atom thick.

Devices

Cameras

The simplest cameras for X-ray powder diffraction consist of a small capillary and either a flat plate detector (originally a piece of X-ray film, now more and more a flat-plate detector or a CCD-camera) or a cylindrical one (originally a piece of film in a cookie-jar, but increasingly bent position sensitive detectors are used). The two types of cameras are known as the Laue and the Debye–Scherrer camera.

In order to ensure complete powder averaging, the capillary is usually spun around its axis.

For neutron diffraction vanadium cylinders are used as sample holders. Vanadium has a negligible absorption and coherent scattering cross section for neutrons and is hence nearly invisible in a powder diffraction experiment. Vanadium does however have a considerable incoherent scattering cross section which may cause problems for more sensitive techniques such as neutron inelastic scattering.

A later development in X-ray cameras is the Guinier camera. It is built around a focusing bent crystal monochromator. The sample is usually placed in the focusing beam, e.g. as a dusting on a piece of sticky tape. A cylindrical piece of film (or electronic multichannel detector) is put on the focusing circle, but the incident beam prevented from reaching the detector to prevent damage from its high intensity.

Diffractometers

Diffractometers can be operated both in transmission and in reflection configurations. The reflection one is more common. The powder sample is filled in a small disc-like container and its surface carefully flattened. The disc is put on one axis of the diffractometer and tilted by an angle θ while a detector (scintillation counter) rotates around it on an arm at twice this angle. This configuration is known under the name Bragg–Brentano theta-2 theta.

Another configuration is the Bragg–Brentano theta-theta configuration in which the sample is stationary while the X-ray tube and the detector are rotated around it. The angle formed between the tube and the detector is 2theta. This configuration is most convenient for loose powders.

Position-sensitive detectors (PSD) and area detectors, which allow collection from multiple angles at once, are becoming more popular on currently supplied instrumentation.

Neutron diffraction

Sources that produce a neutron beam of suitable intensity and speed for diffraction are only available at a small number of research reactors and spallation sources in the world. Angle dispersive (fixed wavelength) instruments typically have a battery of individual detectors arranged in a cylindrical fashion around the sample holder, and can therefore collect scattered intensity simultaneously on a large 2θ range. Time of flight instruments normally have a small range of banks at different scattering angles which collect data at varying resolutions.

X-ray tubes

Laboratory X-ray diffraction equipment relies on the use of an X-ray tube, which is used to produce the X rays. The most commonly used laboratory X-ray tube uses a copper anode, but cobalt and molybdenum are also popular. The wavelength in nm varies for each source. The table below shows these wavelengths, determined by Bearden and quoted in the International Tables for X-ray Crystallography (all values in nm):

Element	Kα (weight average)	Kα2 (strong)	Kα1 (very strong)	Kβ (weak)
Cr	0.229100	0.229361	0.228970	0.208487
Fe	0.193736	0.193998	0.193604	0.175661
Co	0.179026	0.179285	0.178897	0.162079
Cu	0.154184	0.154439	0.154056	0.139222
Mo	0.071073	0.071359	0.070930	0.063229

According to the last re-examination of Holzer et al. (1997), these values are respectively:

Element	Kα2	Kα1	Kβ
Cr	0.2293663	0.2289760	0.2084920
Co	0.1792900	0.1789010	0.1620830
Cu	0.1544426	0.1540598	0.1392250
Mo	0.0713609	0.0709319	0.0632305

Other Sources

In-house applications of X-ray diffraction has always been limited to the relatively few wavelengths shown in the table above. The available choice was much needed because the combination of certain wavelengths and certain elements present in a sample can lead to strong fluorescence which increases the background in the diffraction pattern. A notorious example is the presence of iron in a sample when using copper radiation. In general elements just below the anode element in the period system need to be avoided.

Another limitation is that the intensity of traditional generators is relatively low, requiring lengthy exposure times and precluding any time dependent measurement. The advent of synchrotron sources has drastically changed this picture and caused powder diffraction methods to enter a whole new phase of development. Not only is there a much wider choice of wavelengths available, the high brilliance of the synchrotron radiation makes it possible to observe changes in the pattern during chemical reactions, temperature ramps, changes in pressure and the like.

The tunability of the wavelength also makes it possible to observe anomalous scattering effects when the wavelength is chosen close to the absorption edge of one of the elements of the sample.

Neutron diffraction has never been an in house technique because it requires the availability of an intense neutron beam only available at a nuclear reactor or spallation source. Typically the available neutron flux, and the weak interaction between neutrons and matter, require relative large samples.

Advantages and Disadvantages

Although it is possible to solve crystal structures from powder X-ray data alone, its single crystal analogue is a far more powerful technique for structure determination. This is directly related to the fact that much information is lost by the collapse of the 3D space onto a 1D axis. Nevertheless powder X-ray diffraction is a powerful and useful technique in its own right. It is mostly used to characterize and identify phases, and to refine details of an already known structure, rather than solving unknown structures.

Advantages of the technique are:

- simplicity of sample preparation

- rapidity of measurement
- the ability to analyze mixed phases, e.g. soil samples
- "in situ" structure determination

By contrast growth and mounting of large single crystals is notoriously difficult. In fact there are many materials for which, despite many attempts, it has not proven possible to obtain single crystals. Many materials are readily available with sufficient microcrystallinity for powder diffraction, or samples may be easily ground from larger crystals. In the field of solid-state chemistry that often aims at synthesizing new materials, single crystals thereof are typically not immediately available. Powder diffraction is therefore one of the most powerful methods to identify and characterize new materials in this field.

Particularly for neutron diffraction, which requires larger samples than X-ray diffraction due to a relatively weak scattering cross section, the ability to use large samples can be critical, although newer and more brilliant neutron sources are being built that may change this picture.

Since all possible crystal orientations are measured simultaneously, collection times can be quite short even for small and weakly scattering samples. This is not merely convenient, but can be essential for samples which are unstable either inherently or under X-ray or neutron bombardment, or for time-resolved studies. For the latter it is desirable to have a strong radiation source. The advent of synchrotron radiation and modern neutron sources has therefore done much to revitalize the powder diffraction field because it is now possible to study temperature dependent changes, reaction kinetics and so forth by means of time-resolved powder diffraction.

NEUTRON DIFFRACTION

Neutron diffraction or **elastic neutron scattering** is the application of neutron scattering to the determination of the atomic and/or magnetic structure of a material. A sample to be examined is placed in a beam of thermal or coldneutrons to obtain a diffraction pattern that provides information of the structure of the material. The technique is similar to X-ray diffraction but due to their different scattering properties, neutrons and X-rays provide complementary information: X-Rays are suited for superficial analysis, strong x-rays from synchrotron radiation are suited for shallow depths or thin specimens, while neutrons having high penetration depth are suited for bulk samples.

Instrumental and Sample Requirements

The technique requires a source of neutrons. Neutrons are usually produced in a nuclear reactor or spallation source. At a research reactor, other components are needed, including a crystal monochromators as well as filters to select the desired neutron wavelength. Some parts of the setup may also be movable. At a spallation source, the time of flight technique is used to sort the energies of the incident neutrons (higher energy neutrons are faster), so no monochromator is needed, but rather a series of aperture elements synchronized to filter neutron pulses with the desired wavelength.

The technique is most commonly performed as powder diffraction, which only requires a polycrystalline powder. Single crystal work is also possible, but the crystals must be much larger than those that are used in single-crystal X-ray crystallography. It is common to use crystals that are about 1 mm^3.

Summarizing, the main disadvantage to neutron diffraction is the requirement for a nuclear reactor. For single crystal work, the technique requires relatively large crystals, which are usually challenging to grow. The advantages to the technique are many - sensitivity to light atoms, ability to distinguish isotopes, absence of radiation damage, as well as a penetration depth of several cm

Nuclear Scattering

Like all quantum particles, neutrons can exhibit wave phenomena typically associated with light or sound. Diffraction is one of these phenomena; it occurs when waves encounter obstacles whose size is comparable with the wavelength. If the wavelength of a quantum particle is short enough, atoms or their nuclei can serve as diffraction obstacles. When a beam of neutrons emanating from a reactor is slowed down and selected properly by their speed, their wavelength lies near one angstrom (0.1 nanometer), the typical separation between atoms in a solid material. Such a beam can then be used to perform a diffraction experiment. Impinging on a crystalline sample it will scatter under a limited number of well-defined angles according to the same Bragg's law that describes X-ray diffraction.

Neutrons and X-rays interact with matter differently. X-rays interact primarily with the electron cloud surrounding each atom. The contribution to the diffracted x-ray intensity is therefore larger for atoms with larger atomic number (Z). On the other hand, neutrons interact directly with the nucleus of the atom, and the contribution to the diffracted intensity depends on each isotope; for example, regular hydrogen and deuterium contribute differently. It is also often the case that light (low Z) atoms contribute strongly to the diffracted intensity even in the presence of large Z atoms. The scattering length varies from isotope to isotope rather than linearly with the atomic number. An element like vanadium is a strong scatterer of X-rays, but its nuclei hardly scatter neutrons, which is why it is often used as a container material. Non-magnetic neutron diffraction is directly sensitive to the positions of the nuclei of the atoms.

Unlike X-rays, neutrons scatter mostly from the nuclei of the atoms, which are tiny. Furthermore, there is no need for an atomic form factor to describe the shape of the electron cloud of the atom and the scattering power of an atom does not fall off with the scattering angle as it does for X-rays. Diffractograms therefore can show strong well defined diffraction peaks even at high angles, particularly if the experiment is done at low temperatures. Many neutron sources are equipped with liquid helium cooling systems that allow data collection at temperatures down to 4.2 K. The superb high angle (i.e. high resolution) information means that the atomic positions in the structure can be determined with high precision. On the other hand, Fourier maps (and to a lesser extent difference Fourier maps) derived from neutron data suffer from series termination errors, sometimes so much that the results are meaningless.

Magnetic Scattering

Although neutrons are uncharged, they carry a spin, and therefore interact with magnetic moments, including those arising from the electron cloud around an atom. Neutron diffraction can therefore reveal the microscopic magnetic structure of a material.

Magnetic scattering does require an atomic form factor as it is caused by the much larger electron cloud around the tiny nucleus. The intensity of the magnetic contribution to the diffraction peaks will therefore decrease towards higher angles.

Uses

Neutron diffraction can be used to determine the static structure factor of gases, liquids or amorphous solids. Most experiments, however, aim at the structure of crystalline solids, making neutron diffraction an important tool of crystallography.

Neutron diffraction is closely related to X-ray powder diffraction. In fact the single crystal version of the technique is less commonly used because currently available neutron sources require relatively large samples and large single crystals are hard or impossible to come by for most materials. Future developments, however, may well change this picture. Because the data is typically a 1D powder diffractogram they are usually processed using Rietveld refinement. In fact the latter found its origin in neutron diffraction (at Petten in the Netherlands) and was later extended for use in X-ray diffraction.

One practical application of elastic neutron scattering/diffraction is that the lattice constant of metals and other crystalline materials can be very accurately measured. Together with an accurately aligned micropositioner a map of the lattice constant through the metal can be derived. This can easily be converted to the stress field experienced by the material. This has been used to analyse stresses in aerospace and automotive components to give just two examples. The high penetration depth permits measuring residual stresses in bulk components as crankshafts, pistons, rails, gears. This technique has led to the development of dedicated stress diffractometers, such as the ENGIN-X instrument at the ISIS neutron source.

Neutron diffraction can also be employed to give insight into the 3D structure any material that diffracts.

Hydrogen, Null-Scattering and Contrast Variation

Neutron diffraction can be used to establish the structure of low atomic number materials like proteins and surfactants much more easily with lower flux than at a synchrotron radiation source. This is because some low atomic number materials have a higher cross section for neutron interaction than higher atomic weight materials.

One major advantage of neutron diffraction over X-ray diffraction is that the latter is rather insensitive to the presence of hydrogen (H) in a structure, whereas the nuclei ^{1}H and ^{2}H (i.e. Deuterium, D) are strong scatterers for neutrons. The greater scattering power of protons and deuterons means that the position of hydrogen in a crystal and its thermal motions can be determined with greater precision by neutron diffraction. The structures of metal hydride complexes, e.g., Mg_2FeH_6 have been assessed by neutron diffraction.

The neutron scattering lengths b_H = -3.7406(11) fm and b_D = 6.671(4) fm, for H and D respectively, have opposite sign, which allows the technique to distinguish them. In fact there is a particular isotope ratio for which the contribution of the element would cancel, this is called null-scattering.

It is undesirable to work with the relatively high concentration of H in a sample. The scattering intensity by H-nuclei has a large inelastic component, which creates a large continuous background that is more or less independent of scattering angle. The elastic pattern typically consists of sharp Bragg reflections if the sample is crystalline. They tend to drown in the inelastic background. This is even more serious when the technique is used for the study of liquid structure. Nevertheless, by preparing samples with different isotope ratios it is possible to vary the scattering contrast enough to highlight one element in an otherwise complicated structure. The variation of other elements is possible but usually rather expensive. Hydrogen is inexpensive and particularly interesting because it plays an exceptionally large role in biochemical structures and is difficult to study structurally in other ways.

The first neutron diffraction experiments were carried out in 1945 by Ernest O. Wollan using the Graphite Reactor at Oak Ridge. He was joined shortly thereafter (June 1946) by Clifford Shull, and together they established the basic principles of the technique, and applied it successfully to many different materials, addressing problems like the structure of ice and the microscopic arrangements of magnetic moments in materials. For this achievement Shull was awarded one half of the 1994 Nobel Prize in Physics. Wollan had died in 1984. (The other half of the 1994 Nobel Prize for Physics went to Bert Brockhouse for development of the inelastic scattering technique at the Chalk River facility of AECL. This also involved the invention of the triple axis spectrometer). The delay between the achieved work (1946) and the Nobel Prize awarded to Brockhouse and Shull (1994) brings them close to the delay between the invention by Ernst Ruska of the electron microscope (1933) - also in the field of particle optics - and his own Nobel prize (1986). This in turn is near to the record of 55 years between the discoveries of Peyton Rous and his award of the Nobel Prize in 1966.

REFERENCES

1. Kepler J (1611). Strena seu de Nive Sexangula. Frankfurt: G. Tampach. ISBN 3-321-00021-0.
2. Steno N (1669). De solido intra solidum naturaliter contento dissertationis prodromus. Florentiae.
3. Hessel JFC (1831). Kristallometrie oder Kristallonomie und Kristallographie. Leipzig.
4. Bravais A (1850). "Mémoire sur les systèmes formés par des points distribués regulièrement sur un plan ou dans l'espace. Journal de l'Ecole Polytechnique. ***19***: 1.
5. Shafranovskii I I & Belov N V (1962). Paul Ewald, ed.E. S. Fedorov (PDF). 50 Years of X-Ray Diffraction. Springer: 351. ISBN 90-277-9029-9.
6. Schönflies A (1891). Kristallsysteme und Kristallstruktur. Leipzig.
7. Barlow W (1883).Probable nature of the internal symmetry of crystals. Nature.

29 (738): 186. Bibcode:1883Natur..29..186B. doi:10.1038/029186a0.See also Barlow, William (1883). "Probable Nature of the Internal Symmetry of Crystals". Nature. *29* (739): 205. Bibcode:1883Natur..29..205B. doi:10.1038/029205a0.Sohncke, L. (1884).Probable Nature of the Internal Symmetry of Crystals. Nature. *29* (747): 383. Bibcode:1884Natur..29..383S. doi:10.1038/029383a0.Barlow, WM. (1884). Probable Nature of the Internal Symmetry of Crystals. Nature. *29* (748): 404. Bibcode:1884Natur..29..404B. doi:10.1038/029404b0.

8. Einstein A (1905).Über einen die Erzeugung und Verwandlung des Lichtes betreffenden heuristischen Gesichtspunkt [A Heuristic Model of the Creation and Transformation of Light]. Annalen der Physik (in German). *17* (6): 132. Bibcode:1905AnP...322..132E. doi:10.1002/andp.19053220607.. An English translation is available from Wikisource.
9. Compare: Einstein A (1909).Über die Entwicklung unserer Anschauungen über das Wesen und die Konstitution der Strahlung [The Development of Our Views on the Composition and Essence of Radiation)]. Physikalische Zeitschrift (in German). *10*: 817.. An English translation is available from Wikisource.
10. Pais A (1982). Subtle is the Lord: The Science and the Life of Albert Einstein. Oxford University Press. ISBN 0-19-853907-X.
11. Compton A (1923).A Quantum Theory of the Scattering of X-rays by Light Elements. Phys. Rev. *21* (5): 483. Bibcode:1923PhRv...21..483C. doi:10.1103/PhysRev.21.483.
12. Bragg WH (1907).The nature of Röntgen rays. Transactions of the Royal Society of Science of Australia. *31*: 94.
13. Bragg WH (1908).The nature of γ- and X-rays. Nature. *77* (1995): 270. Bibcode:1908Natur..77..270B. doi:10.1038/077270a0. See also Bragg, W. H. (1908). The Nature of the γ and X-Rays. Nature. *78* (2021): 271. Bibcode:1908Natur..78..271B. doi:10.1038/078271a0.Bragg, W. H. (1908).The Nature of the γ and X-Rays. Nature. *78* (2022): 293. Bibcode:1908Natur..78..293B. doi:10.1038/078293d0. Bragg, W. H. (1908).The Nature of X-Rays. Nature. *78* (2035): 665. Bibcode:1908Natur..78R.665B. doi:10.1038/078665b0.
14. Bragg WH (1910).The consequences of the corpuscular hypothesis of the γ- and X-rays, and the range of β-rays. Phil. Mag. *20* (117): 385. doi:10.1080/14786441008636917.
15. Bragg WH (1912).On the direct or indirect nature of the ionization by X-rays. Phil. Mag. *23* (136): 647. doi:10.1080/14786440408637253.
16. Friedrich W; Knipping P; von Laue M (1912).Interferenz-Erscheinungen bei Röntgenstrahlen. Sitzungsberichte der Mathematisch-Physikalischen Classe der Königlich-Bayerischen Akademie der Wissenschaften zu München. *1912*: 303.
17. von Laue M (1914).Concerning the detection of x-ray interferences (PDF). Nobel Lectures, Physics. 1901–1921. Retrieved 2009-02-18.
18. Dana ES; Ford WE (1932). A Textbook of Mineralogy (fourth ed.). New York: John Wiley & Sons. p. 28.
19. Andre Guinier (1952). X-ray Crystallographic Technology. London: Hilger and

Watts LTD. p. 271.

20. Bragg WL (1912).The Specular Reflexion of X-rays. Nature. ***90*** (2250): 410. Bibcode:1912Natur..90..410B. doi:10.1038/090410b0.
21. Bragg WL (1913).The Diffraction of Short Electromagnetic Waves by a Crystal. Proceedings of the Cambridge Philosophical Society. ***17***: 43.
22. Bragg (1914).Die Reflexion der Röntgenstrahlen. Jahrbuch der Radioaktivität und Elektronik. ***11***: 350.
23. Bragg (1913).The Structure of Some Crystals as Indicated by their Diffraction of X-rays. Proc. R. Soc. Lond. ***A89*** (610): 248–277. Bibcode:1913RSPSA..89..248B. doi:10.1098/rspa.1913.0083. JSTOR 93488.
24. Bragg WL; James RW; Bosanquet CH (1921).The Intensity of Reflexion of X-rays by Rock-Salt. Phil. Mag. ***41*** (243): 309. doi:10.1080/14786442108636225.
25. Bragg WL; James RW; Bosanquet CH (1921).The Intensity of Reflexion of X-rays by Rock-Salt. Part II. Phil. Mag. ***42*** (247): 1. doi:10.1080/14786442108633730.
26. Bragg WL; James RW; Bosanquet CH (1922).The Distribution of Electrons around the Nucleus in the Sodium and Chlorine Atoms. Phil. Mag. ***44*** (261): 433. doi:10.1080/14786440908565188.
27. Bragg WH; Bragg WL (1913).The structure of the diamond. Nature. ***91*** (2283): 557. Bibcode:1913Natur..91..557B. doi:10.1038/091557a0.
28. Bragg WH; Bragg WL (1913).The structure of the diamond. Proc. R. Soc. Lond. ***A89*** (610): 277. Bibcode:1913RSPSA..89..277B. doi:10.1098/rspa.1913.0084.
29. Bragg WL (1914).The Crystalline Structure of Copper. Phil. Mag. ***28*** (165): 355. doi:10.1080/14786440908635219.
30. Bragg WL (1914).The analysis of crystals by the X-ray spectrometer. Proc. R. Soc. Lond. ***A89*** (613): 468. Bibcode:1914RSPSA..89..468B. doi:10.1098/rspa.1914.0015.
31. Bragg WH (1915).The structure of the spinel group of crystals. Phil. Mag. ***30*** (176): 305. doi:10.1080/14786440808635400.
32. Nishikawa S (1915).Structure of some crystals of spinel group. Proc. Tokyo Math. Phys. Soc. ***8***: 199.
33. Vegard L (1916).Results of Crystal Analysis. Phil. Mag. ***32*** (187): 65. doi:10.1080/14786441608635544.
34. Aminoff G (1919).Crystal Structure of Pyrochroite. Stockholm Geol. Fören. Förh. ***41***: 407.
35. Aminoff G (1921).Über die Struktur des Magnesiumhydroxids. Z. Kristallogr. ***56***: 505.
36. Bragg WL (1920).The crystalline structure of zinc oxide. Phil. Mag. ***39*** (234): 647. doi:10.1080/14786440608636079.
37. Debije P, Scherrer P (1916).Interferenz an regellos orientierten Teilchen im Röntgenlicht I. Physikalische Zeitschrift. ***17***: 277.
38. Friedrich W (1913).Eine neue Interferenzerscheinung bei Röntgenstrahlen. Physikalische Zeitschrift. ***14***: 317.

39. Hull AW (1917).A New Method of X-ray Crystal Analysis. Phys. Rev. ***10*** (6): 661. Bibcode:1917PhRv...10..661H. doi:10.1103/PhysRev.10.661.

40. Bernal JD (1924).The Structure of Graphite. Proc. R. Soc. Lond. ***A106*** (740): 749–773. JSTOR 94336.

41. Hassel O; Mack H (1924).Über die Kristallstruktur des Graphits. Zeitschrift für Physik. ***25***: 317. Bibcode:1924ZPhy...25..317H. doi:10.1007/BF01327534.

42. Hull AW (1917).The Crystal Structure of Iron. Phys. Rev. ***9***: 84. Bibcode:1917PhRv....9...83.. doi:10.1103/PhysRev.9.83.

43. Hull AW (1917).The Crystal Structure of Magnesium. PNAS. ***3*** (7): 470. Bibcode:1917PNAS....3..470H. doi:10.1073/pnas.3.7.470.

44. Black, Susan AW (2005).Domesticating the Crystal: Sir Lawrence Bragg and the Aesthetics ofX-ray Analysis. Configurations. ***13*** (2): 257. doi:10.1353/con.2007.0014.

45. From Atoms To Patterns. Wellcome Collection. Archived from the original on September 7, 2013. Retrieved 17 October 2013.

46. Wyckoff RWG; Posnjak E (1921).The Crystal Structure of Ammonium Chloroplatinate. J. Amer. Chem. Soc. ***43*** (11): 2292. doi:10.1021/ja01444a002.

47. Bragg WH (1921).The structure of organic crystals. Proc. R. Soc. Lond. ***34***: 33. Bibcode:1921PPSL...34...33B. doi:10.1088/1478-7814/34/1/306.

48. Lonsdale K (1928).The structure of the benzene ring. Nature. ***122*** (3082): 810. Bibcode:1928Natur.122..810L. doi:10.1038/122810c0.

49. Pauling L. The Nature of the Chemical Bond (3rd ed.). Ithaca, NY: Cornell University Press. ISBN 0-8014-0333-2.

50. Bragg WH (1922).The crystalline structure of anthracene. Proc. R. Soc. Lond. ***35***: 167. Bibcode:1922PPSL...35..167B. doi:10.1088/1478-7814/35/1/320.

51. Powell HM; Ewens RVG (1939).The crystal structure of iron enneacarbonyl. J. Chem. Soc.: 286. doi:10.1039/jr9390000286.

52. Bertrand JA, Cotton, Dollase (1963).The Metal-Metal Bonded, Polynuclear Complex Anion in $CsReCl_4$". J. Amer. Chem. Soc. ***85*** (9): 1349. doi:10.1021/ja00892a029.

53. Robinson WT; Fergusson JE; Penfold BR (1963).Configuration of Anion in $CsReCl_4$". Proceedings of the Chemical Society of London: 116.

54. Cotton FA, Curtis, Harris, Johnson, Lippard, Mague, Robinson, Wood (1964). Mononuclear and Polynuclear Chemistry of Rhenium (III): Its Pronounced Homophilicity. Science. ***145*** (3638): 1305–7. Bibcode:1964Sci...145.1305C. doi:10.1126/science.145.3638.1305. PMID 17802015.

55. Cotton FA, Harris (1965).The Crystal and Molecular Structure of Dipotassium Octachlorodirhenate(III) Dihydrate. Inorganic Chemistry. ***4*** (3): 330. doi:10.1021/ic50025a015.

56. Cotton FA (1965).Metal-Metal Bonding in $[Re_2X_8]^{2-}$ Ions and Other Metal Atom Clusters". Inorganic Chemistry. ***4*** (3): 334. doi:10.1021/ic50025a016.

57. Eberhardt WH; Crawford W, Jr.; Lipscomb WN (1954).The valence structure of the boron hydrides. J. Chem. Phys. ***22*** (6): 989. Bibcode:1954JChPh..22..989E. doi:10.1063/1.1740320.

58. Martin TW; Derewenda ZS (1999).The name is Bond—H bond. Nature Structural Biology. ***6*** (5): 403–6. doi:10.1038/8195. PMID 10331860.

59. Dunitz JD; Orgel LE; Rich A (1956).The crystal structure of ferrocene. Acta Crystallographica. ***9*** (4): 373. doi:10.1107/S0365110X56001091.

60. Seiler P; Dunitz JD (1979).A new interpretation of the disordered crystal structure of ferrocene. Acta Crystallographica B. ***35*** (5): 1068. doi:10.1107/S0567740879005598.

61. Wunderlich JA; Mellor DP (1954).A note on the crystal structure of Zeise's salt. Acta Crystallographica. ***7***: 130. doi:10.1107/S0365110X5400028X.

62. Jarvis JAJ; Kilbourn BT; Owston PG (1970).A re-determination of the crystal and molecular structure of Zeise's salt, $KPtCl_3.C_2H_4.H_2O$. A correction". Acta Crystallographica B. ***26*** (6): 876. doi:10.1107/S056774087000328X.

63. Jarvis JAJ; Kilbourn BT; Owston PG (1971).A re-determination of the crystal and molecular structure of Zeise's salt, $KPtCl_3.C_2H_4.H_2O$". Acta Crystallographica B. ***27*** (2): 366. doi:10.1107/S0567740871002231.

64. Love RA; Koetzle TF; Williams GJB; Andrews LC; Bau R (1975).Neutron diffraction study of the structure of Zeise's salt, $KPtCl_3(C_2H_4).H_2O$". Inorganic Chemistry. ***14*** (11): 2653. doi:10.1021/ic50153a012.

65. Brown, Dwayne (October 30, 2012).NASA Rover's First Soil Studies Help Fingerprint Martian Minerals. NASA. Retrieved October 31, 2012.

66. Westgren A; Phragmén G (1925).X-ray Analysis of the Cu-Zn, Ag-Zn and Au-Zn Alloys. Phil. Mag. ***50***: 311.

67. Bradley AJ; Thewlis J (1926).The structure of γ-Brass. Proc. R. Soc. Lond. ***112*** (762): 678. Bibcode:1926RSPSA.112..678B. doi:10.1098/rspa.1926.0134.

68. Hume-Rothery W (1926).Researches on the Nature, Properties and Conditions of Formation of Intermetallic Compounds (with special Reference to certain Compounds of Tin). Journal of the Institute of Metals. ***35***: 295.

69. Bradley AJ; Gregory CH (1927). The Structure of certain Ternary Alloys. Nature. ***120*** (3027): 678. Bibcode:1927Natur.120..678.. doi:10.1038/120678a0.

70. Westgren A (1932).Zur Chemie der Legierungen. Angewandte Chemie. ***45*** (2): 33. doi:10.1002/ange.19320450202.

71. Bernal JD (1935).The Electron Theory of Metals. Annual Reports on the Progress of Chemistry. ***32***: 181.

72. Pauling L (1923).The Crystal Structure of Magnesium Stannide. J. Amer. Chem. Soc. ***45*** (12): 2777. doi:10.1021/ja01665a001.

73. Pauling L (1929).The Principles Determining the Structure of Complex Ionic Crystals. J. Amer. Chem. Soc. ***51*** (4): 1010. doi:10.1021/ja01379a006.

74. Dickinson RG; Raymond AL (1923).The Crystal Structure of Hexamethylene-

Tetramine. J. Amer. Chem. Soc. ***45***: 22. doi:10.1021/ja01654a003.

75. Müller A (1923).The X-ray Investigation of Fatty Acids. Journal of the Chemical Society (London). ***123***: 2043. doi:10.1039/ct9232302043.

76. Saville WB; Shearer G (1925).An X-ray Investigation of Saturated Aliphatic Ketones. Journal of the Chemical Society (London). ***127***: 591. doi:10.1039/ct9252700591.

77. Bragg WH (1925).The Investigation of thin Films by Means of X-rays. Nature. ***115*** (2886): 266. Bibcode:1925Natur.115..266B. doi:10.1038/115266a0.

78. de Broglie M, Trillat JJ (1925).Sur l'interprétation physique des spectres X d'acides gras. Comptes rendus hebdomadaires des séances de l'Académie des sciences. ***180***: 1485.

79. Trillat JJ (1926).Rayons X et Composeés organiques à longe chaine. Recherches spectrographiques sue leurs structures et leurs orientations. Annales de Physique. ***6***: 5.

80. Caspari WA (1928).Crystallography of the Aliphatic Dicarboxylic Acids. Journal of the Chemical Society (London). ***?***: 3235. doi:10.1039/jr9280003235.

81. Müller A (1928).X-ray Investigation of Long Chain Compounds (n. Hydrocarbons). Proc. R. Soc. Lond. ***120*** (785): 437. Bibcode:1928RSPSA.120..437M. doi:10.1098/rspa.1928.0158.

82. Piper SH (1929).Some Examples of Information Obtainable from the long Spacings of Fatty Acids. Transactions of the Faraday Society. ***25***: 348. doi:10.1039/tf9292500348.

83. Müller A (1929).The Connection between the Zig-Zag Structure of the Hydrocarbon Chain and the Alternation in the Properties of Odd and Even Numbered Chain Compounds. Proc. R. Soc. Lond. ***124*** (794): 317. Bibcode:1929RSPSA.124..317M. doi:10.1098/rspa.1929.0117.

84. Robertson JM (1936).An X-ray Study of the Phthalocyanines, Part II. Journal of the Chemical Society: 1195.

85. Crowfoot Hodgkin D (1935).X-ray Single Crystal Photographs of Insulin. Nature. ***135*** (3415): 591. Bibcode:1935Natur.135..591C. doi:10.1038/135591a0.

86. Kendrew J. C., et al. (1958-03-08).A Three-Dimensional Model of the Myoglobin Molecule Obtained by X-Ray Analysis. Nature. ***181*** (4610): 662–6. Bibcode:1958Natur.181..662K. doi:10.1038/181662a0. PMID 13517261.

87. Table of entries in the PDB, arranged by experimental method.

88. PDB Statistics. RCSB Protein Data Bank. Retrieved 2010-02-09.

89. Scapin G (2006).Structural biology and drug discovery. Curr. Pharm. Des. ***12*** (17): 2087–97. doi:10.2174/138161206777585201. PMID 16796557.

90. Lundstrom K (2006).Structural genomics for membrane proteins. Cell. Mol. Life Sci. ***63*** (22): 2597–607. doi:10.1007/s00018-006-6252-y. PMID 17013556.

91. Lundstrom K (2004).Structural genomics on membrane proteins: mini review. Comb. Chem. High Throughput Screen. ***7*** (5): 431–9. doi:10.2174/1386207043328634.

PMID 15320710.

92. Cryogenic (<20 K) helium cooling mitigates radiation damage to protein crystals" Acta Crystallographica Section D. 2007 63 (4) 486-492

93. J. Claydon, N. Greeves, S. Warren: Organic Chemistry 2nd edition page 45; Oxford University Press 2012

94. Greninger AB (1935).A back-reflection Laue method for determining crystal orientation. Zeitschrift für Kristallographie. ***91***: 424.

95. An analogous diffraction pattern may be observed by shining a laser pointer on a compact disc or DVD; the periodic spacing of the CD tracks corresponds to the periodic arrangement of atoms in a crystal.

96. Miao, J., Charalambous, P., Kirz, J., & Sayre, D. (1999).Extending the methodology of X-ray crystallography to allow imaging of micrometre-sized non-crystalline specimens.Nature, 400(6742), 342.

97. Harp, JM; Timm, DE; Bunick, GJ (1998).Macromolecular crystal annealing: overcoming increased mosaicity associated with cryocrystallography. Acta Crystallographica D. ***54*** (Pt 4): 622–8. doi:10.1107/S0907444997019008. PMID 9761858.

98. Harp, JM; Hanson, BL; Timm, DE; Bunick, GJ (1999).Macromolecular crystal annealing: evaluation of techniques and variables. Acta Crystallographica D. ***55*** (Pt 7): 1329–34. doi:10.1107/S0907444999005442. PMID 10393299.

99. Hanson, BL; Harp, JM; Bunick, GJ (2003).The well-tempered protein crystal: annealing macromolecular crystals. Methods in enzymology. Methods in Enzymology. ***368***: 217–35. doi:10.1016/S0076-6879(03)68012-2. ISBN 978-0-12-182271-2. PMID 14674276.

100. Geerlof A, et al. (2006).The impact of protein characterization in structural proteomics. Acta Crystallographica D. ***62*** (Pt 10): 1125–36. doi:10.1107/S0907444906030307. PMID 17001090.

101. Chernov AA (2003).Protein crystals and their growth. J. Struct. Biol. ***142*** (1): 3–21. doi:10.1016/S1047-8477(03)00034-0. PMID 12718915.

102. Rupp B; Wang J (2004).Predictive models for protein crystallization. Methods. ***34*** (3): 390–407. doi:10.1016/j.ymeth.2004.03.031. PMID 15325656.

103. Chayen NE (2005).Methods for separating nucleation and growth in protein crystallization. Prog. Biophys. Mol. Biol. ***88*** (3): 329–37. doi:10.1016/j.pbiomolbio.2004.07.007. PMID 15652248.

104. Stock D; Perisic O; Lowe J (2005).Robotic nanolitre protein crystallisation at the MRC Laboratory of Molecular Biology. Prog Biophys Mol Biol. ***88*** (3): 311–27. doi:10.1016/j.pbiomolbio.2004.07.009. PMID 15652247.

105. Jeruzalmi D (2006).First analysis of macromolecular crystals: biochemistry and x-ray diffraction. Methods Mol. Biol. ***364***: 43–62. doi:10.1385/1-59745-266-1:43. ISBN 1-59745-266-1. PMID 17172760.

106. Helliwell JR (2005).Protein crystal perfection and its application. Acta Crystallographica D. ***61*** (Pt 6): 793–8. doi:10.1107/S0907444905001368. PMID

15930642.

107. Garman, E. F.; Schneider, T. R. (1997).Macromolecular Cryocrystallography. Journal of Applied Crystallography. ***30*** (3): 211. doi:10.1107/S0021889897002677.

108. Schlichting, I; Miao, J (2012).Emerging opportunities in structural biology with X-ray free-electron lasers. Current Opinion in Structural Biology. ***22*** (5): 613–26. doi:10.1016/j.sbi.2012.07.015. PMC 3495068. PMID 22922042.

109. Neutze, R; Wouts, R; Van Der Spoel, D; Weckert, E; Hajdu, J (2000).Potential for biomolecular imaging with femtosecond X-ray pulses. Nature. ***406*** (6797): 752–7. Bibcode:2000Natur.406..752N. doi:10.1038/35021099. PMID 10963603.

110. Liu, W; Wacker, D; Gati, C; Han, G. W.; James, D; Wang, D; Nelson, G; Weierstall, U; Katritch, V; Barty, A; Zatsepin, N. A.; Li, D; Messerschmidt, M; Boutet, S; Williams, G. J.; Koglin, J. E.; Seibert, M. M.; Wang, C; Shah, S. T.; Basu, S; Fromme, R; Kupitz, C; Rendek, K. N.; Grotjohann, I; Fromme, P; Kirian, R. A.; Beyerlein, K. R.; White, T. A.; Chapman, H. N.; et al. (2013).Serial femtosecond crystallography of G protein-coupled receptors. Science. ***342*** (6165): 1521–4. Bibcode:2013Sci...342.1521L. doi:10.1126/science.1244142. PMC 3902108. PMID 24357322.

111. Ravelli RB; Garman EF (2006).Radiation damage in macromolecular cryocrystallography. Curr. Opin. Struct. Biol. ***16*** (5): 624–9. doi:10.1016/j.sbi.2006.08.001. PMID 16938450.

112. Powell HR (1999).The Rossmann Fourier autoindexing algorithm in MOSFLM. Acta Crystallographica D. ***55*** (Pt 10): 1690–5. doi:10.1107/S0907444999009506. PMID 10531518.

113. Hauptman H (1997).Phasing methods for protein crystallography. Curr. Opin. Struct. Biol. ***7*** (5): 672–80. doi:10.1016/S0959-440X(97)80077-2. PMID 9345626.

114. Us6n I; Sheldrick GM (1999).Advances in direct methods for protein crystallography. Curr. Opin. Struct. Biol. ***9*** (5): 643–8. doi:10.1016/S0959-440X(99)00020-2. PMID 10508770.

115. Taylor G (2003).The phase problem. Acta Crystallographica D. ***59*** (11): 1881. doi:10.1107/S0907444903017815.

116. Ealick SE (2000).Advances in multiple wavelength anomalous diffraction crystallography. Current Opinion in Chemical Biology. ***4*** (5): 495–9. doi:10.1016/S1367-5931(00)00122-8. PMID 11006535.

117. Parkin, G.,Bond-stretch isomerism in transition metal complexes: a reevaluation of crystallographic data, Chem. Rev. 1993, volume 93, 887-911. doi:10.1021/cr00019a003

118. Patterson AL (1935).A Direct Method for the Determination of the Components of Interatomic Distances in Crystals. Zeitschrift für Kristallographie. ***90***: 517. doi:10.1524/zkri.1935.90.1.517.

119. The Nobel Prize in Physics 1914. Nobel Foundation. Retrieved 2008-10-09.

120. The Nobel Prize in Physics 1915. Nobel Foundation. Retrieved 2008-10-09.

121. The Nobel Prize in Chemistry 1962. Nobelprize.org. Retrieved 2008-10-06.

122. The Nobel Prize in Physiology or Medicine 1962. Nobel Foundation. Retrieved 2007-07-28.
123. The Nobel Prize in Chemistry 1964. Nobelprize.org. Retrieved 2008-10-06.
124. The Nobel Prize in Chemistry 1972. Nobelprize.org. Retrieved 2008-10-06.
125. The Nobel Prize in Chemistry 1976. Nobelprize.org. Retrieved 2008-10-06.
126. The Nobel Prize in Chemistry 1985. Nobelprize.org. Retrieved 2008-10-06.
127. The Nobel Prize in Chemistry 1988. Nobelprize.org. Retrieved 2008-10-06.
128. The Nobel Prize in Chemistry 1997. Nobelprize.org. Retrieved 2008-10-06.
129. The Nobel Prize in Chemistry 2003. Nobelprize.org. Retrieved 2008-10-06.
130. The Nobel Prize in Chemistry 2006. Nobelprize.org. Retrieved 2008-10-06.
131. The Nobel Prize in Chemistry 2009. Nobelprize.org. Retrieved 2009-10-07.
132. The Nobel Prize in Chemistry 2012. Nobelprize.org. Retrieved 2012-10-13.

CHAPTER 3

Solid

Chapter Contents

- **Introduction**
- **Microscopic Description**
- **Classes of Solids**

A solid is a sample of matter that retains its shape and density when not confined. The adjective solid describes the state, or condition, of matter having this property. The atom s or molecule s of matter in the solid state are generally compressed as tightly as the repulsive forces among them will allow. Some solids, called crystalline solids, tend to fracture along defined surfaces that have a characteristic shape depending on the arrangement of, and the forces among, the atoms or molecules in the sample. Other solids, known as amorphous solids, lack any apparent crystalline structure.

INTRODUCTION

Solid is one of the four fundamental states of matter (the others being liquid, gas, and plasma). It is characterized by structural rigidity and resistance to changes of shape or volume. Unlike a liquid, a solid object does not flow to take on the shape of its container, nor does it expand to fill the entire volume available to it like a gas does. The atoms in a solid are tightly bound to each other, either in a regular geometric lattice (crystalline solids, which include metals and ordinary ice) or irregularly (an amorphous solid such as common window glass).

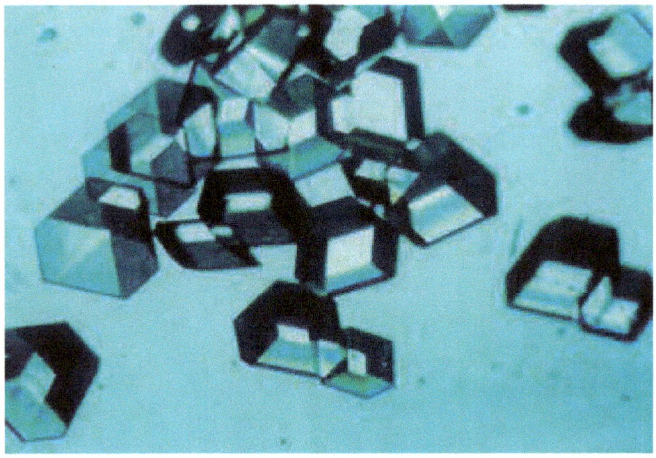

Single crystalline form of solid insulin.

The branch of physics that deals with solids is called solid-state physics, and is the main branch of condensed matter physics (which also includes liquids). Materials science is primarily concerned with the physical and chemical properties of solids. Solid-state chemistry is especially concerned with the synthesis of novel materials, as well as the science of identification and chemical composition.

MICROSCOPIC DESCRIPTION

Model of closely packed atoms within a crystalline solid.

The atoms, molecules or ions which make up solids may be arranged in an orderly repeating pattern, or irregularly. Materials whose constituents are arranged in a regular pattern are known as crystals. In some cases, the regular ordering can continue unbroken over a large scale, for example diamonds, where each diamond is a single crystal. Solid objects that are large enough to see and handle are rarely composed of a single crystal, but instead are made of a large number of single crystals, known as crystallites, whose size can vary from a few nanometers to several meters. Such materials are called polycrystalline. Almost all common metals, and many ceramics, are polycrystalline.

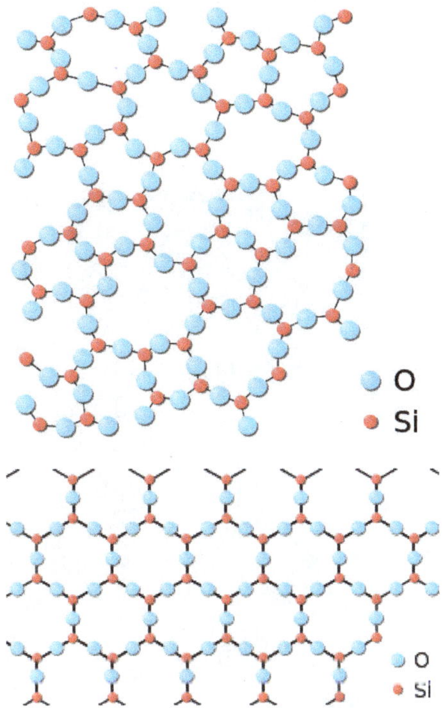

Schematic representation of a random-network glassy form (left) and ordered crystalline lattice (right) of identical chemical composition.

In other materials, there is no long-range order in the position of the atoms. These solids are known as amorphous solids; examples include polystyrene and glass.

Whether a solid is crystalline or amorphous depends on the material involved, and the conditions in which it was formed. Solids which are formed by slow cooling will tend to be crystalline, while solids which are frozen rapidly are more likely to be amorphous. Likewise, the specific crystal structure adopted by a crystalline solid depends on the material involved and on how it was formed.

While many common objects, such as an ice cube or a coin, are chemically identical throughout, many other common materials comprise a number of different substances packed together. For example, a typical rock is an aggregate of several different minerals and mineraloids, with no specific chemical composition. Wood is a natural organic material consisting primarily of cellulose fibers embedded in a matrix of organic lignin. In materials science, composites of more than one constituent material can be designed to have desired properties.

CLASSES OF SOLIDS

The forces between the atoms in a solid can take a variety of forms. For example, a crystal of sodium chloride (common salt) is made up of ionic sodium and chlorine, which are held together by ionic bonds. In diamond or silicon, the atoms share electrons and form covalent bonds. In metals, electrons are shared in metallic bonding. Some solids, particularly most organic compounds, are held together with van der Waals forces resulting from the polarization of the electronic charge cloud on each molecule. The dissimilarities between the types of solid result from the differences between their bonding.

Metals

The pinnacle of New York's Chrysler Building, the world's tallest steel-supported brick building, is clad with stainless steel.

Metals typically are strong, dense, and good conductors of both electricity and heat. The bulk of the elements in the periodic table, those to the left of a diagonal line drawn from boron to polonium, are metals. Mixtures of two or more elements in which the major component is a metal are known as alloys.

People have been using metals for a variety of purposes since prehistoric times. The strength and reliability of metals has led to their widespread use in construction of buildings and other structures, as well as in most vehicles, many appliances and tools, pipes, road signs and railroad tracks. Iron and aluminium are the two most commonly used structural metals, and they are also the most abundant metals in the Earth›s crust. Iron is most commonly used in the form of an alloy, steel, which contains up to 2.1% carbon, making it much harder than pure iron.

Because metals are good conductors of electricity, they are valuable in electrical appliances and for carrying an electric current over long distances with little energy loss or dissipation. Thus, electrical power grids rely on metal cables to distribute electricity. Home electrical systems, for example, are wired with copper for its good conducting properties and easy machinability. The high thermal conductivity of most metals also makes them useful for stovetop cooking utensils.

The study of metallic elements and their alloys makes up a significant portion of the fields of solid-state chemistry, physics, materials science and engineering.

Metallic solids are held together by a high density of shared, delocalized electrons, known as metallic bonding. In a metal, atoms readily lose their outermost (valence) electrons, forming positive ions. The free electrons are spread over the entire solid, which is held together firmly by electrostatic interactions between the ions and the electron cloud. The large number of free electrons gives metals their high values of electrical and thermal conductivity. The free electrons also prevent transmission of visible light, making metals opaque, shiny and lustrous.

More advanced models of metal properties consider the effect of the positive ions cores on the delocalised electrons. As most metals have crystalline structure, those ions are usually arranged into a periodic lattice. Mathematically, the potential of the ion cores can be treated by various models, the simplest being the nearly free electron model.

Minerals

A collection of various minerals.

Minerals are naturally occurring solids formed through various geological processes under high pressures. To be classified as a true mineral, a substance must have a crystal structure with uniform physical properties throughout. Minerals range in composition from pure elements and simple salts to very complex silicates with thousands of known forms. In contrast, a rock sample is a random aggregate of minerals and/or mineraloids, and has no specific chemical composition. The vast majority of the rocks of the Earth›s crust consist of quartz (crystalline SiO_2), feldspar, mica, chlorite, kaolin, calcite, epidote, olivine, augite, hornblende, magnetite, hematite, limonite and a few other minerals. Some minerals, like quartz, mica or feldspar are common, while others have been found in only a few locations worldwide. The largest group of minerals by far is the silicates (most rocks are ≥95% silicates), which are composed largely of silicon and oxygen, with the addition of ions of aluminium, magnesium, iron, calcium and other metals.

Ceramics

Si_3N_4 ceramic bearing parts

Ceramic solids are composed of inorganic compounds, usually oxides of chemical elements. They are chemically inert, and often are capable of withstanding chemical erosion that occurs in an acidic or caustic environment. Ceramics generally can withstand high temperatures ranging from 1000 to 1600 °C (1800 to 3000 °F). Exceptions include non-oxide inorganic materials, such as nitrides, borides and carbides.

Traditional ceramic raw materials include clay minerals such as kaolinite, more recent materials include aluminium oxide (alumina). The modern ceramic materials, which are classified as advanced ceramics, include silicon carbide and tungsten carbide. Both are valued for their abrasion resistance, and hence find use in such applications as the wear plates of crushing equipment in mining operations.

Most ceramic materials, such as alumina and its compounds, are formed from fine powders, yielding a fine grained polycrystalline microstructure which is filled with light scattering centers comparable to the wavelength of visible light. Thus, they are generally opaque materials, as opposed to transparent materials. Recent nanoscale (e.g. sol-gel) technology has, however, made possible the production of polycrystalline transparent ceramics such as transparent alumina and alumina compounds for such applications as high-power lasers. Advanced ceramics are also used in the medicine, electrical and electronics industries.

Ceramic engineering is the science and technology of creating solid-state ceramic materials, parts and devices. This is done either by the action of heat, or, at lower temperatures, using precipitation reactions from chemical solutions. The term includes the purification of raw materials, the study and production of the chemical compounds concerned, their formation into components, and the study of their structure, composition and properties.

Mechanically speaking, ceramic materials are brittle, hard, strong in compression and weak in shearing and tension. Brittle materials may exhibit significant tensile strength by supporting a static load. Toughness indicates how much energy a material can absorb before mechanical failure, while fracture toughness (denoted K_{Ic}) describes the ability of a material with inherent microstructural flaws to resist fracture via crack growth and propagation. If a material has a large value of fracture toughness, the basic principles of fracture mechanics suggest that it will most likely undergo ductile fracture. Brittle fracture is very characteristic of most ceramic and glass-ceramic materials which typically exhibit low (and inconsistent) values of K_{Ic}.

For an example of applications of ceramics, the extreme hardness of zirconia is utilized in the manufacture of knife blades, as well as other industrial cutting tools. Ceramics such as alumina, boron carbide and silicon carbide have been used in bulletproof vests to repel large-caliber rifle fire. Silicon nitride parts are used in ceramic ball bearings, where their high hardness makes them wear resistant. In general, ceramics are also chemically resistant and can be used in wet environments where steel bearings would be susceptible to oxidation (or rust).

As another example of ceramic applications, in the early 1980s, Toyota researched production of an adiabatic ceramic engine with an operating temperature of over 6000 °F (3300 °C). Ceramic engines do not require a cooling system and hence allow a major weight reduction and therefore greater fuel efficiency. In a conventional metallic engine, much of the energy released from the fuel must be dissipated as waste heat in order to prevent a meltdown of the metallic parts. Work is also being done in developing ceramic parts for gas turbine engines. Turbine engines made with ceramics could operate more efficiently, giving aircraft greater range and payload for a set amount of fuel. However, such engines are not in production because the manufacturing of ceramic parts in the sufficient precision and durability is difficult and costly. Processing methods often result in a wide distribution of microscopic flaws which frequently play a detrimental role in the sintering process, resulting in the proliferation of cracks, and ultimate mechanical failure.

Glass Ceramics

A high strength glass-ceramic cooktop with negligible thermal expansion.

Glass-ceramic materials share many properties with both non-crystalline glasses and crystalline ceramics. They are formed as a glass, and then partially crystallized by heat treatment, producing both amorphous and crystalline phases so that crystalline grains are embedded within a non-crystalline intergranular phase.

Glass-ceramics are used to make cookware (originally known by the brand name CorningWare) and stovetops which have both high resistance to thermal shock and extremely low permeability to liquids. The negative coefficient of thermal expansion of the crystalline ceramic phase can be balanced with the positive coefficient of the glassy phase. At a certain point (~70% crystalline) the glass-ceramic has a net coefficient of thermal expansion close to zero. This type of glass-ceramic exhibits excellent mechanical properties and can sustain repeated and quick temperature changes up to 1000 °C.

Glass ceramics may also occur naturally when lightning strikes the crystalline (e.g. quartz) grains found in most beach sand. In this case, the extreme and immediate heat of the lightning (~2500 °C) creates hollow, branching rootlike structures called fulgurite via fusion.

Organic Solids

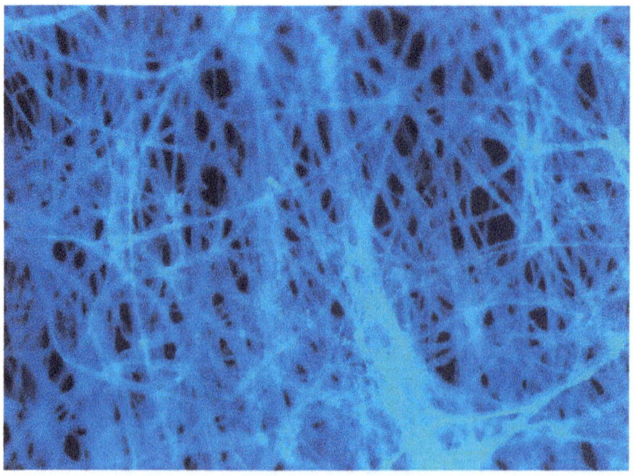

The individual wood pulp fibers in this sample are around 10 μm in diameter.

Organic chemistry studies the structure, properties, composition, reactions, and preparation by synthesis (or other means) of chemical compounds of carbon and hydrogen, which may contain any number of other elements such as nitrogen, oxygen and the halogens: fluorine, chlorine, bromine and iodine. Some organic compounds may also contain the elements phosphorus or sulfur. Examples of organic solids include wood, paraffin wax, naphthalene and a wide variety of polymers and plastics.

Wood

Wood is a natural organic material consisting primarily of cellulose fibers embedded in a matrix of lignin. Regarding mechanical properties, the fibers are strong in tension, and the lignin matrix resists compression. Thus wood has been an important construction material since humans began building shelters and using boats. Wood to be used for construction work is commonly known as *lumber* or *timber*. In construction, wood is not only a structural material, but is also used to form the mould for concrete.

Wood-based materials are also extensively used for packaging (e.g. cardboard) and paper which are both created from the refined pulp. The chemical pulping processes use a combination of high temperature and alkaline (kraft) or acidic (sulfite) chemicals to break the chemical bonds of the lignin before burning it out.

Polymers

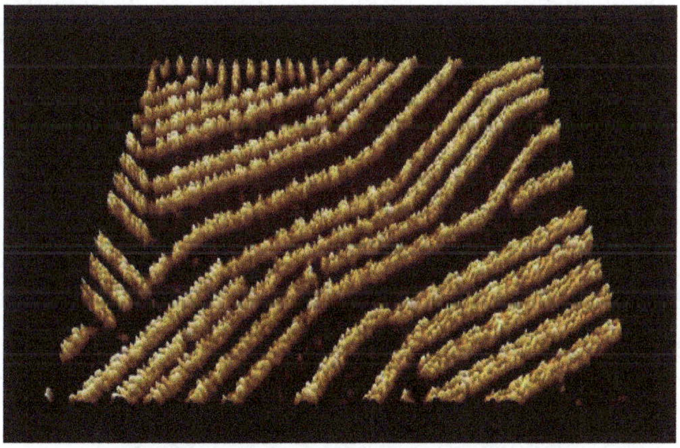

STM image of self-assembled supramolecular chains of the organic semiconductor quinacridone on graphite.

One important property of carbon in organic chemistry is that it can form certain compounds, the individual molecules of which are capable of attaching themselves to one another, thereby forming a chain or a network. The process is called polymerization and the chains or networks polymers, while the source compound is a monomer. Two main groups of polymers exist: those artificially manufactured are referred to as industrial polymers or synthetic polymers (plastics) and those naturally occurring as biopolymers.

Monomers can have various chemical substituents, or functional groups, which can affect the chemical properties of organic compounds, such as solubility and chemical reactivity, as well as the physical properties, such as hardness, density, mechanical or tensile strength, abrasion resistance, heat resistance, transparency, color, etc.. In proteins, these differences give the polymer the ability to adopt a biologically active conformation in preference to others (see self-assembly).

Household items made of various kinds of plastic.

People have been using natural organic polymers for centuries in the form of waxes and shellac which is classified as a thermoplastic polymer. A plant polymer named cellulose provided the tensile strength for natural fibers and ropes, and by the early 19th century natural rubber was in widespread use. Polymers are the raw materials (the resins) used to make what are commonly called plastics. Plastics are the final product, created after one or more polymers or additives have been added to a resin during processing, which is then shaped into a final form. Polymers which have been around, and which are in current widespread use, include carbon-based polyethylene, polypropylene, polyvinyl chloride, polystyrene, nylons, polyesters, acrylics, polyurethane, and polycarbonates, and silicon-based silicones. Plastics are generally classified as commodity, specialty and engineering plastics.

Composite Materials

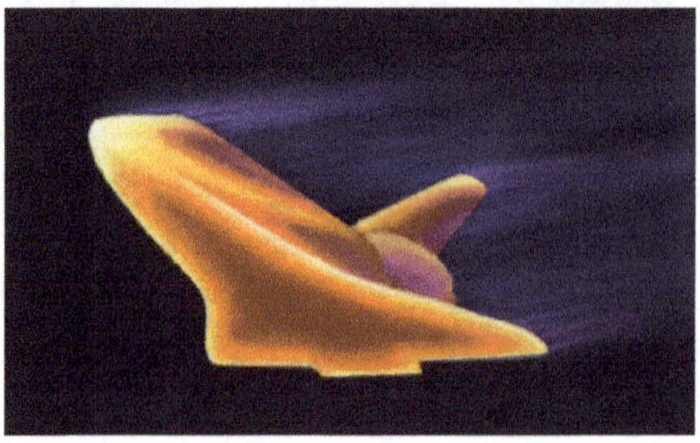

Simulation of the outside of the Space Shuttle as it heats up to over 1500 °C during re-entry

A cloth of woven carbon fiber filaments, a common element in composite materials

Composite materials contain two or more macroscopic phases, one of which is often ceramic. For example, a continuous matrix, and a dispersed phase of ceramic particles or fibers.

Applications of composite materials range from structural elements such as steel-reinforced concrete, to the thermally insulative tiles which play a key and integral role in NASA›s Space Shuttle thermal protection system which is used to protect the surface of the shuttle from the heat of re-entry into the Earth's atmosphere. One example is Reinforced Carbon-Carbon (RCC), the light gray material which withstands reentry temperatures up to 1510 °C (2750 °F) and protects the nose cap and leading edges of Space Shuttle's wings. RCC is a laminated composite material made from graphite rayon cloth and impregnated with a phenolic resin. After curing at high temperature in an autoclave, the laminate is pyrolized to convert the resin to carbon, impregnated with furfural alcohol in a vacuum chamber, and cured/pyrolized to convert the furfural alcohol to carbon. In order to provide oxidation resistance for reuse capability, the outer layers of the RCC are converted to silicon carbide.

Domestic examples of composites can be seen in the plastic casings of television sets, cell-phones and so on. These plastic casings are usually a composite made up of a thermoplastic matrix such as acrylonitrile butadiene styrene (ABS) in which calcium carbonate chalk, talc, glass fibers or carbon fibers have been added for strength, bulk, or electro-static dispersion. These additions may be referred to as reinforcing fibers, or dispersants, depending on their purpose.

Thus, the matrix material surrounds and supports the reinforcement materials by maintaining their relative positions. The reinforcements impart their special mechanical and physical properties to enhance the matrix properties. A synergism produces material properties unavailable from the individual constituent materials, while the wide variety of matrix and strengthening materials provides the designer with the choice of an optimum combination.

Semiconductors

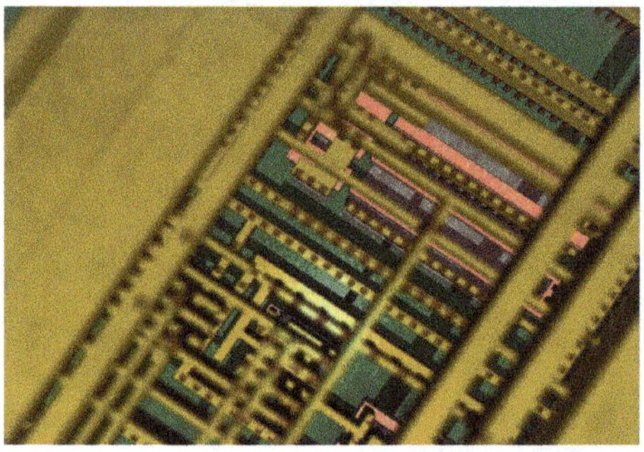

Semiconductor chip on crystalline silicon substrate.

Semiconductors are materials that have an electrical resistivity (and conductivity) between that of metallic conductors and non-metallic insulators. They can be found in the periodic table moving diagonally downward right from boron. They separate the electrical conductors (or metals, to the left) from the insulators (to the right).

Devices made from semiconductor materials are the foundation of modern electronics, including radio, computers, telephones, etc. Semiconductor devices include the transistor, solar cells, diodes and integrated circuits. Solar photovoltaic panels are large semiconductor devices that directly convert light into electrical energy.

In a metallic conductor, current is carried by the flow of electrons, but in semiconductors, current can be carried either by electrons or by the positively charged holes in the electronic band structure of the material. Common semiconductor materials include silicon, germanium and gallium arsenide.

Nanomaterials

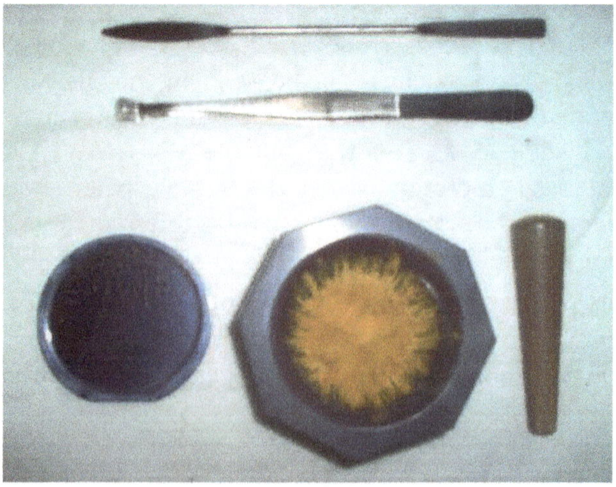

Bulk silicon (left) and silicon nanopowder (right).

Many traditional solids exhibit different properties when they shrink to nanometer sizes. For example, nanoparticles of usually yellow gold and gray silicon are red in color; gold nanoparticles melt at much lower temperatures (~300 °C for 2.5 nm size) than the gold slabs (1064 °C); and metallic nanowires are much stronger than the corresponding bulk metals. The high surface area of nanoparticles makes them extremely attractive for certain applications in the field of energy. For example, platinum metals may provide improvements as automotive fuel catalysts, as well as proton exchange membrane (PEM) fuel cells. Also, ceramic oxides (or cermets) of lanthanum, cerium, manganese and nickel are now being developed as solid oxide fuel cells (SOFC). Lithium, lithium–titanate and tantalum nanoparticles are being applied in lithium ion batteries. Silicon nanoparticles have been shown to dramatically expand the storage capacity of lithium ion batteries during the expansion/contraction cycle. Silicon nanowires cycle without significant degradation and present the potential for use in batteries with greatly expanded storage times. Silicon nanoparticles are also being used in new forms of solar energy cells. Thin film deposition of silicon quantum dots on the polycrystalline silicon substrate of a photovoltaic (solar) cell increases voltage output as much as 60% by fluorescing the incoming light prior to capture. Here again, surface area of the nanoparticles (and thin films) plays a critical role in maximizing the amount of absorbed radiation.

Biomaterials

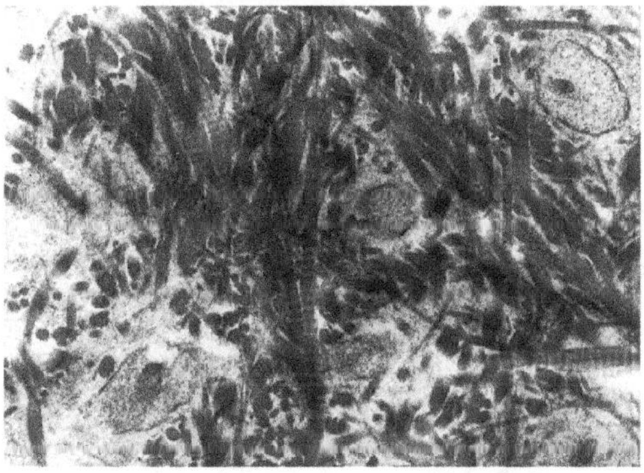

Collagen fibers of woven bone.

Many natural (or biological) materials are complex composites with remarkable mechanical properties. These complex structures, which have risen from hundreds of million years of evolution, are inspiring materials scientists in the design of novel materials. Their defining characteristics include structural hierarchy, multifunctionality and self-healing capability. Self-organization is also a fundamental feature of many biological materials and the manner by which the structures are assembled from the molecular level up. Thus, self-assembly is emerging as a new strategy in the chemical synthesis of high performance biomaterials.

Physical properties

Physical properties of elements and compounds which provide conclusive evidence of chemical composition include odor, color, volume, density (mass per unit volume), melting point, boiling point, heat capacity, physical form and shape at room temperature (solid, liquid or gas; cubic, trigonal crystals, etc.), hardness, porosity, index of refraction and many others. This section discusses some physical properties of materials in the solid state.

Mechanical

Granite rock formation in the Chilean Patagonia. Like most inorganic minerals formed by oxidation in the Earth's atmosphere, granite consists primarily of crystalline silica SiO_2 and alumina Al_2O_3.

The mechanical properties of materials describe characteristics such as their strength and resistance to deformation. For example, steel beams are used in construction because of their high strength, meaning that they neither break nor bend significantly under the applied load.

Mechanical properties include elasticity and plasticity, tensile strength, compressive strength, shear strength, fracture toughness, ductility (low in brittle materials), and indentation hardness. Solid mechanics is the study of the behavior of solid matter under external actions such as external forces and temperature changes.

A solid does not exhibit macroscopic flow, as fluids do. Any degree of departure from its original shape is called deformation. The proportion of deformation to original size is called strain. If the applied stress is sufficiently low, almost all solid materials behave in such a way that the strain is directly proportional to the stress (Hooke›s law). The coefficient of the proportion is called the modulus of elasticity or Young›s modulus. This region of deformation is known as the linearly elastic region. Three models can describe how a solid responds to an applied stress:

- Elasticity – When an applied stress is removed, the material returns to its undeformed state.
- Viscoelasticity – These are materials that behave elastically, but also have damping. When the applied stress is removed, work has to be done against the damping effects and is converted to heat within the material. This results in a hysteresis loop in the stress–strain curve. This implies that the mechanical response has a time-dependence.
- Plasticity – Materials that behave elastically generally do so when the applied stress is less than a yield value. When the stress is greater than the yield stress, the material behaves plastically and does not return to its previous state. That is, irreversible plastic deformation (or viscous flow) occurs after yield which is permanent.

Many materials become weaker at high temperatures. Materials which retain their strength at high temperatures, called refractory materials, are useful for many purposes. For example, glass-ceramics have become extremely useful for countertop cooking, as they exhibit excellent mechanical properties and can sustain repeated and quick temperature changes up to 1000 °C. In the aerospace industry, high performance materials used in the design of aircraft and/or spacecraft exteriors must have a high resistance to thermal shock. Thus, synthetic fibers spun out of organic polymers and polymer/ceramic/metal composite materials and fiber-reinforced polymers are now being designed with this purpose in mind.

Thermal

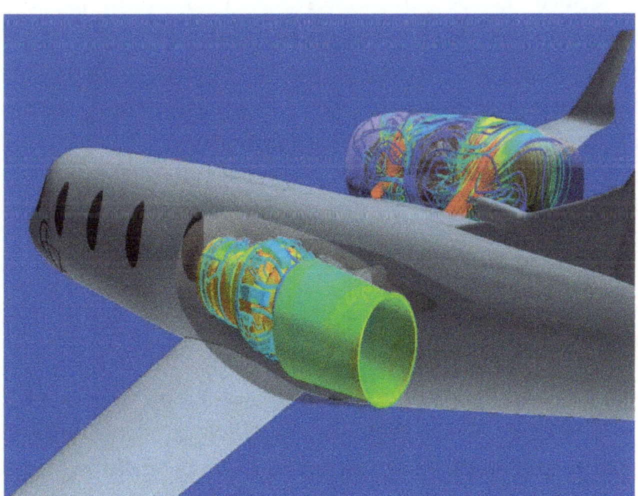

Normal modes of atomic vibration in a crystalline solid.

Because solids have thermal energy, their atoms vibrate about fixed mean positions within the ordered (or disordered) lattice. The spectrum of lattice vibrations in a crystalline or glassy network provides the foundation for the kinetic theory of solids. This motion occurs at the atomic level, and thus cannot be observed or detected without highly specialized equipment, such as that used in spectroscopy.

Thermal properties of solids include thermal conductivity, which is the property of a material that indicates its ability to conduct heat. Solids also have a specific heat capacity, which is the capacity of a material to store energy in the form of heat (or thermal lattice vibrations).

Electrical

Electrical properties include conductivity, resistance, impedance and capacitance. Electrical conductors such as metals and alloys are contrasted with electrical insulators such as glasses and ceramics. Semiconductors behave somewhere in between. Whereas conductivity in metals is caused by electrons, both electrons and holes contribute to current in semiconductors. Alternatively, ions support electric current in ionic conductors.

Many materials also exhibit superconductivity at low temperatures; they include metallic elements such as tin and aluminium, various metallic alloys, some heavily doped semiconductors, and certain ceramics. The electrical resistivity of most electrical (metallic) conductors generally decreases gradually as the temperature is lowered, but remains finite. In a superconductor however, the resistance drops abruptly to zero when the material is cooled below its critical temperature. An electric current flowing in a loop of superconducting wire can persist indefinitely with no power source.

A dielectric, or electrical insulator, is a substance that is highly resistant to the flow of electric current. A dielectric, such as plastic, tends to concentrate an applied electric field within itself which property is used in capacitors. A capacitor is an electrical device that can store energy in the electric field between a pair of closely spaced conductors (called 'plates'). When voltage is applied to the capacitor, electric charges of equal magnitude, but opposite polarity, build up on each plate. Capacitors are used in electrical circuits as energy-storage devices, as well as in electronic filters to differentiate between high-frequency and low-frequency signals.

Electro-mechanical

Piezoelectricity is the ability of crystals to generate a voltage in response to an applied mechanical stress. The piezoelectric effect is reversible in that piezoelectric crystals, when subjected to an externally applied voltage, can change shape by a small amount. Polymer materials like rubber, wool, hair, wood fiber, and silk often behave as electrets. For example, the polymer polyvinylidene fluoride (PVDF) exhibits a piezoelectric response several times larger than the traditional piezoelectric material quartz (crystalline SiO_2). The deformation (~0.1%) lends itself to useful technical applications such as high-voltage sources, loudspeakers, lasers, as well as chemical, biological, and acousto-optic sensors and/or transducers.

Optical

Materials can transmit (e.g. glass) or reflect (e.g. metals) visible light.

Many materials will transmit some wavelengths while blocking others. For example, window glass is transparent to visible light, but much less so to most of the frequencies of ultraviolet light that cause sunburn. This property is used for frequency-selective optical filters, which can alter the color of incident light.

For some purposes, both the optical and mechanical properties of a material can be of interest. For example, the sensors on an infrared homing ("heat-seeking") missile must be protected by a cover which is transparent to infrared radiation. The current material of choice for high-speed infrared-guided missile domes is single-crystal sapphire. The optical transmission of sapphire does not actually extend to cover the entire mid-infrared range (3–5 μm), but starts to drop off at wavelengths greater than approximately 4.5 μm at room temperature. While the strength of sapphire is better than that of other available mid-range infrared dome materials at room temperature, it weakens above 600 °C. A long-standing trade-off exists between optical bandpass and mechanical durability; new materials such as transparent ceramics or optical nanocomposites may provide improved performance.

Guided lightwave transmission involves the field of fiber optics and the ability of certain glasses to transmit, simultaneously and with low loss of intensity, a range of frequencies (multi-mode optical waveguides) with little interference between them. Optical waveguides are used as components in integrated optical circuits or as the transmission medium in optical communication systems.

Opto-Electronic

A solar cell or photovoltaic cell is a device that converts light energy into electrical energy. Fundamentally, the device needs to fulfill only two functions: photo-generation of charge carriers (electrons and holes) in a light-absorbing material, and separation of the charge carriers to a conductive contact that will transmit the electricity (simply put, carrying electrons off through a metal contact into an external circuit). This conversion is called the photoelectric effect, and the field of research related to solar cells is known as photovoltaics.

Solar cells have many applications. They have long been used in situations where electrical power from the grid is unavailable, such as in remote area power systems, Earth-orbiting satellites and space probes, handheld calculators, wrist watches, remote radiotelephones and water pumping applications. More recently, they are starting to be used in assemblies of solar modules (photovoltaic arrays) connected to the electricity grid through an inverter, that is not to act as a sole supply but as an additional electricity source.

All solar cells require a light absorbing material contained within the cell structure to absorb photons and generate electrons via the photovoltaic effect. The materials used in solar cells tend to have the property of preferentially absorbing the wavelengths of solar light that reach the earth surface. However, some solar cells are optimized for light absorption beyond Earth's atmosphere as well.

REFERENCES

1. Mortimer, Charles E. (1975). Chemistry: A Conceptual Approach (3rd ed.). New York:: D. Van Nostrad Company. ISBN 0-442-25545-4.
2. Buffat, Ph.; Borel, J.-P. (1976). Size effect on the melting temperature of gold particles. Physical Review A. **13** *(6): 2287. Bibcode:1976PhRvA..13.2287B. doi:10.1103/PhysRevA.13.2287.*
3. Walter H. Kohl (1995). Handbook of materials and techniques for vacuum devices. Springer. pp. 164–167. ISBN 1-56396-387-6.
4. Shpak, Anatoly P.; Kotrechko, Sergiy O.; Mazilova, Tatjana I; Mikhailovskij, Igor M (2009). Inherent tensile strength of molybdenum nanocrystals. Science and Technology of Advanced Materials. **10** *(4): 045004. Bibcode:2009STAdM..10d5004S. doi:10.1088/1468-6996/10/4/045004. PMC 5090266.*

CHAPTER 4

Ionic Conductivity

Chapter Contents

- **Introduction**
- **Silver Iodide**
- **Fast Ion Conductor**
- **Microporous Material**
- **Mesoporous Material**

Ionic conductivity is the ability of a material to conduct electric current. The conductivity (σ), is expressed as the inverse of resistivity (ρ) i.e., $\sigma = 1/\rho$ and its unit is Siemens. Specific conductivity is the reciprocal of the specific resistance of a solution measured between two electrodes 1 cm^2 in area and 1 cm apart. Conductivity is measured by placing two electrodes (with opposite electrical charge) in the electrolyte. For a known electrical current, the voltage drop across the electrodes reveals the electrolyte's resistance.

INTRODUCTION

Ionic conduction (denoted by λ-lambda) is the movement of an ion from one site to another through defects in the crystal lattice of a solid or aqueous solution.

Ionic conduction is one mechanism of current. In solids, ions typically occupy fixed positions in the crystal lattice and do not move. However, ionic conduction can occur, especially as the temperature increases

Ionic conduction in solids has been a subject of interest since the beginning of the 19th century. Michael Faraday established in 1839 that the laws of electrolysis are also obeyed in ionic solids like lead(II) fluoride (PbF_2) and silver sulfide (Ag_2S).

SILVER IODIDE

In 1921, Tubandt et al. found that solid silver iodide (AgI) had extraordinary high ionic conductivity. At temperatures above 147 °C, AgI changes into a phase that has an ionic conductivity of ~ 1 –1 cm^{-1}, similar to that of its liquid phase. This high temperature phase of AgI was the first superionic conductor ever discovered. The highly conductive phase of AgI is now known as alpha-AgI. It was shown that a sublattice cationic disorder takes place in alpha-AgI. The liquid-like state of Ag^+ ions, as proposed by Strock (1934, 1936) and later reinforced by others (Geller, 1977; Funke, 1976), consists of a cubic unit cell of iodide ions (I^-), in which a total of 42 sites (6 octahedral, 12 tetragonal and 24 trigonal bipyramidal) are available for 2 Ag^+ions, as shown in the Figure 1. O' Keeffe and Hyde (1976) have argued that this phase transition in AgI is dramatic and powerful, nothing less than the melting and have also shown that the entropy change at the superionic transition is comparable to its value at the melting. Thus, in the -phase, I^- ions form a body-centered cubic lattice and the Ag^+ ions are distributed in such a way that 42 crystallographic equivalent interstices are available for the two Ag^+ ions.

Alpha phase crystals of various materials like Ag_2S, Ag_2Se, Ag_2Te, etc. were soon discovered (Tubandt, 1932). By the early 1930s, it was demonstrated that these fast ionically conducting solids could be treated entirely the same as aqueouselectrolytes from the viewpoint of chemical reactions and thermodynamics, hence these materials were labeled *solid electrolytes*.

FAST ION CONDUCTOR

In materials science, **fast ion conductors** are solids in which ions are highly mobile. These materials are important in the area of solid-state ionics, and are also known as **solid electrolytes** and **superionic conductors**. These materials are useful in batteries and various sensors. Fast ion conductors are used primarily in solid oxide fuel cells. As solid electrolytes they allow the movement of ions without the need for a liquid or soft membrane separating the electrodes. The phenomenon relies on the hopping of ions through an otherwise rigid crystal structure.

Mechanism

Fast ion conductors are intermediate in nature between crystalline solids which possess a regular structure with immobile ions, and liquid electrolytes which have no regular structure and fully mobile ions. Solid electrolytes find use in all solid-state supercapacitors, batteries, and fuel cells, and in various kinds of chemical sensors.

Classification

In solid electrolytes (glasses or crystals), the ionic conductivity Ω_i can be any value, but it should be much larger than the electronic one. Usually, solids where Ω_i is on the order of 0.0001 to 0.1 Ohm^{-1} cm^{-1} (300 K) are called superionic conductors.

Proton Conductors

Proton conductors are a special class of solid electrolytes, where hydrogen ions act as charge carriers.

Superionic Conductors

Unlike conventional solid electrolytes and superionic conductors.

Superionic conductors where Ω_i is more than 0.1 Ohm^{-1} cm^{-1} (300 K) and the activation energy for ion transport E_i is small (about 0.1 eV), are called advanced superionic conductors. The most famous example of advanced superionic conductor-solid electrolyte is $RbAg_4I_5$ where $\Omega_i > 0.25$ Ohm^{-1} cm^{-1} and $\Omega_e \sim 10^{-9}$ Ohm^{-1} cm^{-1} at 300 K. The Hall (drift) ionic mobility in $RbAg_4I_5$ is about 2×10^{-4} $cm^2/(V\bullet s)$ at room temperatures. The Ω_e – Ω_i systematic diagram distinguishing the different types of solid-state ionic conductors is given in the figure.

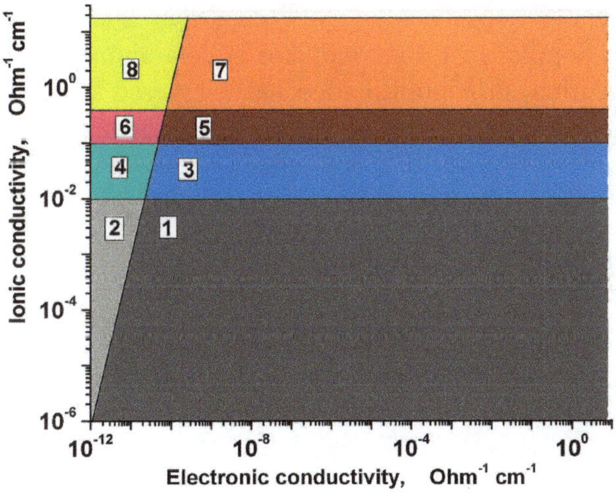

Classification of solid-state ionic conductors by the *lg* (electronic conductivity, Ω_e) – *lg* (ionic conductivity, Ω_i) diagram. 2, 4, and 6 – known solid electrolytes (SEs), materials with $\Omega_i >> \Omega_e$; 1, 3, and 5 – known mixed ion-electron conductors; 3 and 4 – superionic conductors (SICs), i.e. materials with $\Omega_i > 0.001$ $Ohm^{-1}cm^{-1}$, Ω_e – arbitrary value; 4 – SIC and simultaneously SE, $\Omega_i > 0.001$ $Ohm^{-1}cm^{-1}$, $\Omega_i >> \Omega_e$; 5 and 6 – advanced superionic conductors (AdSICs), where $\Omega_i > 10^{-1}$ $Ohm^{-1}cm^{-1}$ (300 K), energy activation E_i about 0.1 eV, Ω_e – arbitrary value; 6 – AdSIC and simultaneously SE, $\Omega_i > 10^{-1}$ $Ohm^{-1}cm^{-1}$, Ei about 0.1 eV, $\Omega_i >> \Omega_e$; 7 and 8 – hypothetical AdSIC with $Ei \approx k_BT \approx 0.03$ eV (300 K); 8 – hypothetical AdSIC and simultaneously SE.

No clear examples have been described as yet, of fast ion conductors in the hypothetical advanced superionic conductors class (areas 7 and 8 in the classification plot). However, in crystal structure of several superionic conductors, e.g. in the minerals of the pearceite-polybasite group, the large structural fragments with activation energy of ion transport $Ei < k_B T$ (300 K) had been discovered in 2006.

Examples

Zirconia-Based Materials

A common solid electrolyte is yttria-stabilized zirconia, YSZ. This material is prepared by doping Y_2O_3 into ZrO_2. Oxide ions typically migrate only slowly in solid Y_2O_3 and in ZrO_2, but in YSZ, the conductivity of oxide increases dramatically. These materials are used to allow oxide to move through the solid in certain kinds of fuel cells. Zirconium dioxide can also be doped with calcium oxide to give an oxide conductor that is used in oxygen sensors in automobile controls. Upon doping only a few percent, the diffusion constant of oxide increases by a factor of ~1000.

Other conductive ceramics function as ion conductors. One example is NASICON, ($Na_3Zr_2Si_2PO_{12}$), a sodium super-ionic conductor

beta-Alumina

Another example of a popular fast ion conductor is beta-alumina solid electrolyte. Unlike the usual forms of alumina, this modification has a layered structure with open galleries separated by pillars. Sodium ions (Na_+) migrate through this material readily since the oxide framework provides an ionophilic, non-reducible medium. This material is considered as the sodium ion conductor for the sodium–sulfur battery.

Fluoride ion conductors

Lanthanum trifluoride (LaF_3) is conductive for F^- ions, used in some ion selective electrodes. Beta-lead fluoride exhibits a continuous growth of conductivity on heating. This property was first discovered by Michael Faraday.

Iodides

A textbook example of a fast ion conductor is silver iodide (AgI). Upon heating the solid to 146 °C, this material adopts the alpha-polymorph. In this form, the iodide ions form a rigid cubic framework, and the Ag+ centers are molten. The electrical conductivity of the solid increases by 4000x. Similar behavior is observed for copper(I) iodide (CuI), rubidium silver iodide ($RbAgI_2$), and Ag_2HgI_4.

Other Inorganic materials

- Silver sulfide, conductive for Ag^+ ions, used in some ion selective electrodes
- Lead(II) chloride, conductive at higher temperatures
- Some perovskite ceramics – strontium titanate, strontium stannate – conductive for O^{2-} ions
- $Zr(HPO_4)_2.nH_2O$ – conductive for H^+ ions
- $UO_2HPO_4.4H_2O$ – conductive for H^+ ions

Organic materials

- Many gels, such polyacrylamides, agar, etc. are fast ion conductors
- A salt dissolved in a polymer – e.g. lithium perchlorate in polyethylene oxide
- Polyelectrolytes and Ionomers – e.g. Nafion, a H^+ conductor

The important case of fast ionic conduction is one in a surface space-charge layer of ionic crystals. Such conduction was first predicted by Kurt Lehovec. As a space-charge layer has nanometer thickness, the effect is directly related to nanoionics (nanoionics-I). Lehovec›s effect is used as a basis for developing nanomaterials for portable lithium batteries and fuel cells.

MICROPOROUS MATERIAL

A **microporous material** is a material containing pores with diameters less than 2 nm. Examples of microporous materials include zeolites and metal-organic frameworks.

Porous materials are classified into several kinds by their size. The recommendations of a panel convened by the International Union of Pure and Applied Chemistry (IUPAC) are:

- microporous materials have pore diameters of less than 2 nm
- mesoporous materials have pore diameters between 2 nm and 50 nm
- macroporous materials have pore diameters of greater than 50 nm.

Micropores may be defined differently in other contexts. For example, in the context of porous aggregations such as soil, micropores are defined as cavities with sizes less than 30 μm.

Uses in Laboratories

Microporous materials are often used in laboratory environments to facilitate contaminant-free exchange of gases. Mold spores, bacteria, and other airborne contaminants will become trapped, while allowing gases to pass through the material. This allows for a sterile environment in the contained area.

Use in Medicine

Albupore microporous surgical tape.

Microporous adhesive tape is a surgical tape used to hold wound dressings and bandages in place, introduced in 1959 by 3M with the trade name *Micropore*. It can be used to hold gauze padding over small wounds, usually as a temporary measure until a suitable dressing is applied. The Steri-Strip was derived from Microporous surgical tape.

Other uses

Microporous tape is used by some professional extreme yo-yoers to wrap around their fingers and prevent string burn or irritation.

Rock climbers use microporous tape to wrap their hands in 'tape gloves', a means of protecting the skin from rock abrasion when jamming hands into cracks as a means of ascent (crack climbing, as opposed to face climbing - gripping holds on the face of the rock).

Microporous media used in large format printing applications normally with a pigment based ink to maintain colour balance and life expectancy of the resultant printed image.

Microporous tape is also used by some film and TV sound recordists to affix small radio microphones to actors' skin.

Microporous material is also used as high performance insulation material used from home applications up to metal furnaces requiring material that can withstand more than 1000 Celsius.

MESOPOROUS MATERIAL

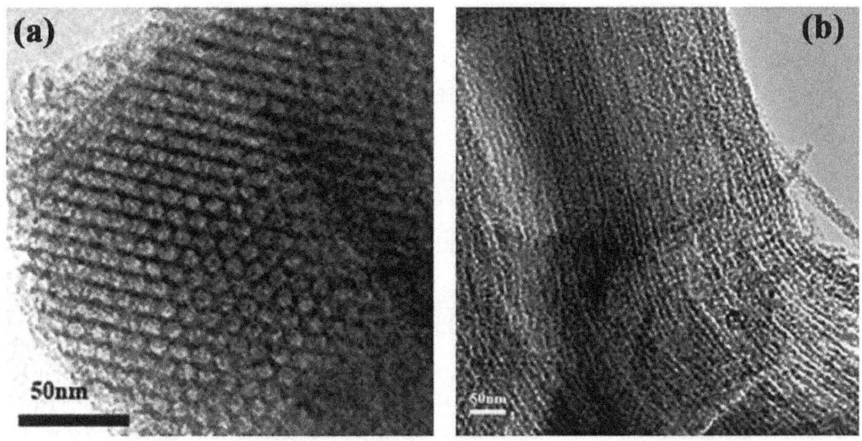

Electron microscopy images of nitrogen-containing ordered mesoporous carbon (N-OMC) taken (a) along and (b) perpendicular to the channel direction.

A **mesoporous material** is a material containing pores with diameters between 2 and 50 nm.

Porous materials are classified into several kinds by their size. According to IUPAC notation, microporous materials have pore diameters of less than 2 nm and macroporous materials have pore diameters of greater than 50 nm; the mesoporous category thus lies in the middle.

Typical mesoporous materials include some kinds of silica and alumina that have similarly-sized fine mesopores. Mesoporous oxides of niobium, tantalum, titanium, zirconium, cerium and tin have also been reported. According to the IUPAC, a mesoporous material can be disordered or ordered in a mesostructure.

A procedure for producing mesoporous materials (silica) was patented around 1970, and methods based on the Stöber process from 1968 were still in use in 2015. It went almost unnoticed and was reproduced in 1997. Mesoporous silica nanoparticles (MSNs) were independently synthesized in 1990 by researchers in Japan. They were later produced also at Mobil Corporation laboratories and named Mobil Crystalline Materials, or MCM-41.

Since then, research in this field has steadily grown. Notable examples of prospective applications are catalysis, sorption, gas sensing, ion exchange, optics, and photovoltaics.

Mesopores may be defined differently in other contexts. For example, in the context of porous aggregations such as soil, mesopores are defined as cavities with sizes in the range 30 μm–75 μm.

REFERENCES

1. Guo, M.; Wang, H.; Huang, D.; Han, Z.; Li, Q.; Wang, X.; Chen, J. (2014). Amperometric catechol biosensor based on laccase immobilized on nitrogen-doped ordered mesoporous carbon (N-OMC)/PVA matrix . Science and Technology of Advanced Materials. 15 (3): 035005. doi:10.1088/1468-6996/15/3/035005. PMC 090526. PMID 27877681.

2. Rouquerol, J.; Avnir, D.; Fairbridge, C. W.; Everett, D. H.; Haynes, J. M.; Pernicone, N.; Ramsay, J. D. F.; Sing, K. S. W.; Unger, K. K. (1994). Recommendations for the characterization of porous solids (Technical Report) . Pure and Applied Chemistry. 66 (8). doi:10.1351/pac199466081739.

3. Chiola, V.; Ritsko, J. E. and Vanderpool, C. D. Process for producing low-bulk density silica. Application No. US 3556725D A filed on 26-Feb-1969; Publication No. US 3556725 A published on 19-Jan-1971

4. Porous silica particles containing a crystallized phase and method Application No. US 3493341D A filed on 23-Jan-1967; Publication No. US 3493341 A published on 03-Feb-1970

5. Process for producing silica in the form of hollow spheres ; Application No. US 342525 A filed on 04-Feb-1964; Publication No. US 3383172 A published on 14-May-1968

6. Stöber, Werner; Fink, Arthur; Bohn, Ernst (1968). Controlled growth of monodisperse silica spheres in the micron size range . Journal of Colloid and Interface Science. 26 (1): 62–69. doi:10.1016/0021-9797(68)90272-5.

7. Kicklebick, Guido (2015). Nanoparticles and Composites . In Levy, David; Zayat, Marcos. The Sol-Gel Handbook: Synthesis, Characterization and Applications. 3. John Wiley & Sons. pp. 227–244. ISBN 9783527334865.

8. Xu, Ruren; Pang, Wenqin & Yu, Jihong (2007). Chemistry of zeolites and related porous materials: synthesis and structure. Wiley-Interscience. p. 472. ISBN 0-470-82233-3.

9. Direnzo, F; Cambon, H; Dutartre, R (1997). A 28-year-old synthesis of micelle-templated mesoporous silica . Microporous Materials. 10 (4–6): 283. doi:10.1016/S0927-6513(97)00028-X.

10. Yanagisawa, Tsuneo; Shimizu, Toshio; Kuroda, Kazuyuki; Kato, Chuzo (1990). The preparation of alkyltrimethylammonium-kanemite complexes and their conversion to microporous materials . Bulletin of the Chemical Society of Japan. 63 (4): 988. doi:10.1246/bcsj.63.988.

11. Beck, J. S.; Vartuli, J. C.; Roth, W. J.; Leonowicz, M. E.; Kresge, C. T.; Schmitt, K. D.; Chu, C. T. W.; Olson, D. H.; Sheppard, E. W. (1992). A new family of mesoporous molecular sieves prepared with liquid crystal templates . Journal of the American Chemical Society. 114 (27): 10834. doi:10.1021/ja00053a020.

12. Trewyn, B. G.; Slowing, I. I.; Giri, S.; Chen, H. T.; Lin, V. S. -Y. (2007). Synthesis and Functionalization of a Mesoporous Silica Nanoparticle Based

on the Sol–Gel Process and Applications in Controlled Release . Accounts of Chemical Research. 40 (9): 846–53. doi:10.1021/ar600032u. PMID 17645305.

13. Soil Science Glossary Terms Committee (2008). Glossary of Soil Science Terms 2008. Madison, WI: Soil Science Society of America. ISBN 978-0-89118-851-3.

CHAPTER 5

Magnetism

Chapter Contents

Magnetism is one aspect of the combined electromagnetic force. It refers to physical phenomena arising from the force caused by magnets, objects that produce fields that attract or repel other objects. A magnetic field exerts a force on particles in the field due to the Lorentz force, according to Georgia State University's HyperPhysics website. The motion of electrically charged particles gives rise to magnetism. The force acting on an electrically charged particle in a magnetic field depends on the magnitude of the charge, the velocity of the particle, and the strength of the magnetic field.

All materials experience magnetism, some more strongly than others. Permanent magnets, made from materials such as iron, experience the strongest effects, known as ferromagnetism. With rare exception, this is the only form of magnetism strong enough to be felt by people.

A magnetic quadrupole.

Magnetism is a class of physical phenomena that are mediated by magnetic fields. Electric currents and the magnetic moments of elementary particles give rise to a magnetic field, which acts on other currents and magnetic moments. The most familiar effects occur in ferromagnetic materials, which are strongly attracted by magnetic fields and can be magnetized to become permanent magnets, producing magnetic fields themselves. Only a few substances are ferromagnetic; the most common ones are iron, nickel and cobalt and their alloys. The prefix *ferro-* refers to iron, because permanent magnetism was first observed in lodestone, a form of natural iron ore called magnetite, Fe_3O_4.

Although ferromagnetism is responsible for most of the effects of magnetism encountered in everyday life, all other materials are influenced to some extent by a magnetic field, by several other types of magnetism. Paramagnetic substances such as aluminum and oxygen are weakly attracted to an applied magnetic field; diamagnetic substances such as copper and carbon are weakly repelled; while antiferromagnetic materials such as chromium and spin glasses have a more complex relationship with a magnetic field. The force of a magnet on paramagnetic, diamagnetic, antiferromagnetic materials is usually too weak to be felt, and can be detected only by laboratory instruments, so in everyday life these substances are often described as non-magnetic.

The magnetic state (or magnetic phase) of a material depends on temperature and other variables such as pressure and the applied magnetic field. A material may exhibit more than one form of magnetism as these variables change.

HISTORY

Lodestone, a natural magnet, attracting iron nails. Ancient humans discovered the property of magnetism from lodestone.

An illustration from Gilbert's 1600 *Die Magnete* showing one of the earliest methods of making a magnet. A blacksmith holds a piece of red-hot iron in a north-south direction and hammers it as it cools. The magnetic field of the Earth aligns the domains, leaving the iron a weak magnet.

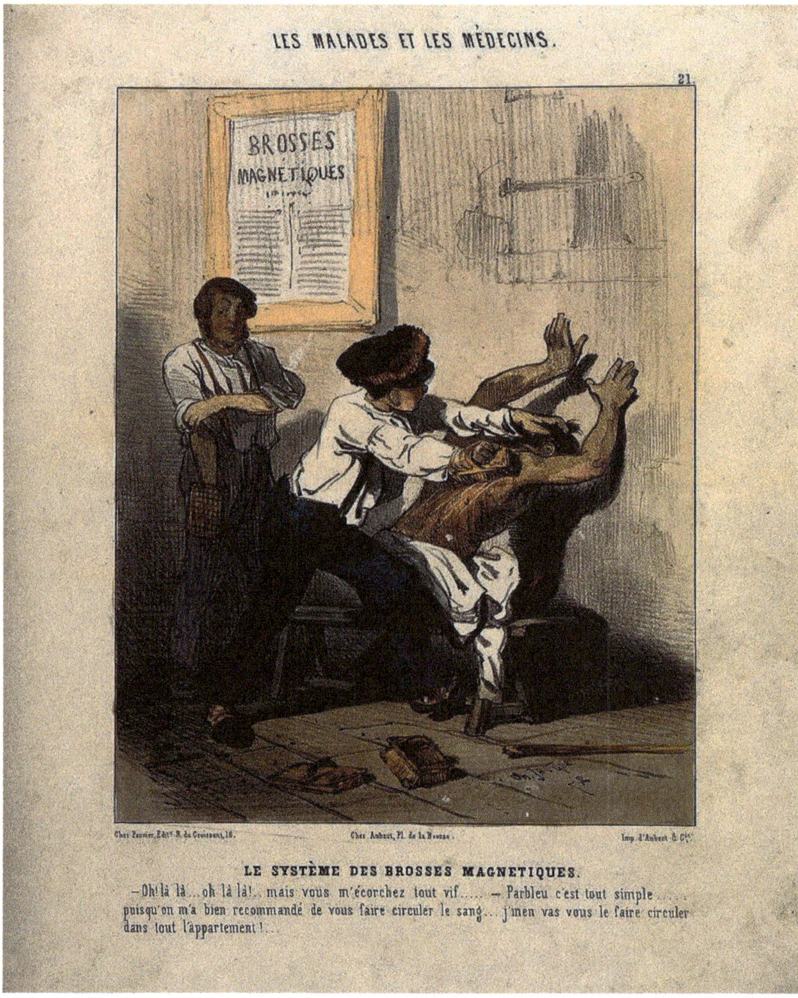

Drawing of a medical treatment using magnetic brushes. Charles Jacque 1843, France.

Magnetism was first discovered in the ancient world, when people noticed that lodestones, naturally magnetized pieces of the mineral magnetite, could attract iron. The word *magnet* comes from the Greek term for lodestone, "magnítis líthos (μαγνήτης λίθος), which means a stone from the region of Magnesia. In ancient Greece, Aristotle attributed the first of what could be called a scientific discussion of magnetism to the philosopher Thales of Miletus, who lived from about 625 BC to about 545 BC. Around the same time, in ancient India, the Indian surgeon Sushruta was the first to make use of the magnet for surgical purposes.

In ancient China, the earliest literary reference to magnetism lies in a 4th-century BC book named after its author, *The Master of Demon Valley*. The 2nd-century BC annals, *Lüshi Chunqiu*, also notes: "The lodestone makes iron approach, or it attracts it." The earliest mention of the attraction of a needle is in a 1st-century work *Lunheng* (*Balanced Inquiries*): "A lodestone attracts a needle." The 11th-century Chinese scientist Shen Kuo was the first person to write – in the *Dream Pool Essays* – of the magnetic needle compass and that it improved the accuracy of navigation by employing the astronomical concept of true north. By the 12th century the Chinese were known to use the lodestone compass for navigation. They sculpted a directional spoon from lodestone in such a way that the handle of the spoon always pointed south.

Alexander Neckam, by 1187, was the first in Europe to describe the compass and its use for navigation. In 1269, Peter Peregrinus de Maricourt wrote the *Epistola de magnete*, the first extant treatise describing the properties of magnets. In 1282, the properties of magnets and the dry compass were discussed by Al-Ashraf, a Yemeni physicist, astronomer, and geographer.

In 1600, William Gilbert published his *De Magnete, Magneticisque Corporibus, et de Magno Magnete Tellure* (*On the Magnet and Magnetic Bodies, and on the Great Magnet the Earth*). In this work he describes many of his experiments with his model earth called the terrella. From his experiments, he concluded that the Earth was itself magnetic and that this was the reason compasses pointed north (previously, some believed that it was the pole star (Polaris) or a large magnetic island on the north pole that attracted the compass).

An understanding of the relationship between electricity and magnetism began in 1819 with work by Hans Christian Ørsted, a professor at the University of Copenhagen, who discovered by the accidental twitching of a compass needle near a wire that an electric current could create a magnetic field. This landmark experiment is known as Ørsted's Experiment. Several other experiments followed, with André-Marie Ampère, who in 1820 discovered that the magnetic field circulating in a closed-path was related to the current flowing through the perimeter of the path; Carl Friedrich Gauss; Jean-Baptiste Biot and Félix Savart, both of whom in 1820 came up with the Biot–Savart law giving an equation for the magnetic field from a current-carrying wire; Michael Faraday, who in 1831 found that a time-varying magnetic flux through a loop of wire induced a voltage, and others finding further links between magnetism and electricity. James Clerk Maxwell synthesized and expanded these insights into Maxwell›s equations, unifying electricity, magnetism, and optics into the field of electromagnetism. In 1905, Einstein used these laws in motivating his theory of special relativity, requiring that the laws held true in all inertial reference frames.

Electromagnetism has continued to develop into the 21st century, being incorporated into the more fundamental theories of gauge theory, quantum electrodynamics, electroweak theory, and finally the standard model.

Sources of Magnetism

Magnetism, at its root, arises from two sources:

- Electric current (see *Electron magnetic moment*).
- Spin magnetic moments of elementary particles. The magnetic moments of the nuclei of atoms are typically thousands of times smaller than the electrons› magnetic moments, so they are negligible in the context of the magnetization of materials. Nuclear magnetic moments are nevertheless very important in other contexts, particularly in nuclear magnetic resonance (NMR) and magnetic resonance imaging (MRI).

Ordinarily, the enormous number of electrons in a material are arranged such that their magnetic moments (both orbital and intrinsic) cancel out. This is due, to some extent, to electrons combining into pairs with opposite intrinsic magnetic moments as a result of the Pauli exclusion principle (see *electron configuration*), or combining into filled subshells with zero net orbital motion. In both cases, the electron arrangement is so as to exactly cancel the magnetic moments from each electron. Moreover, even when the electron configuration *is* such that there are unpaired electrons and/or non-filled subshells, it is often the case that the various electrons in the solid will contribute magnetic moments that point in different, random directions, so that the material will not be magnetic.

Sometimes, either spontaneously, or owing to an applied external magnetic field—each of the electron magnetic moments will be, on average, lined up. A suitable material can then produce a strong net magnetic field.

The magnetic behavior of a material depends on its structure, particularly its electron configuration, for the reasons mentioned above, and also on the temperature. At high temperatures, random thermal motion makes it more difficult for the electrons to maintain alignment.

MATERIALS

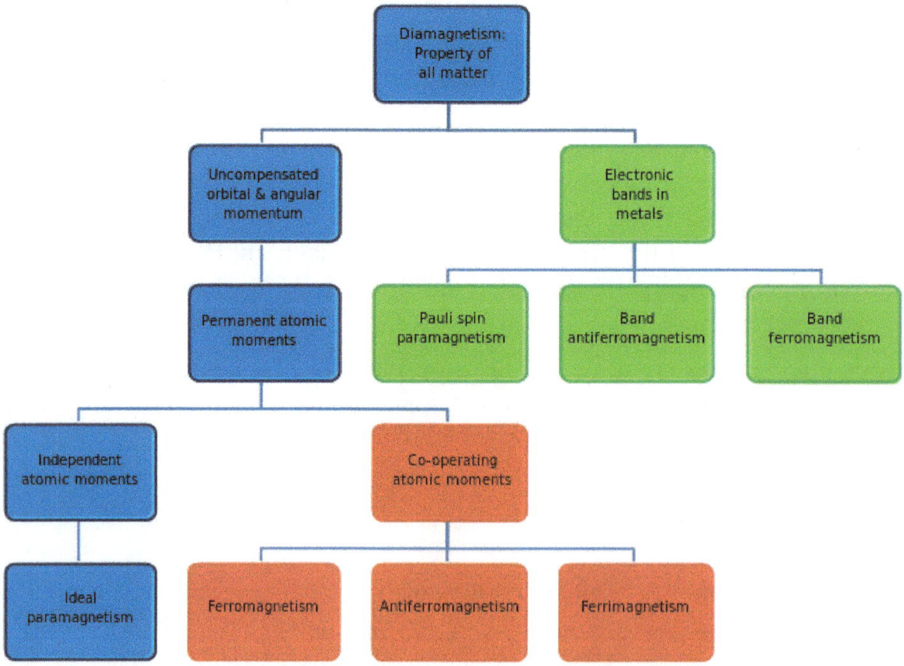

Hierarchy of types of magnetism.

Diamagnetism

Diamagnetism appears in all materials, and is the tendency of a material to oppose an applied magnetic field, and therefore, to be repelled by a magnetic field. However, in a material with paramagnetic properties (that is, with a tendency to enhance an external magnetic field), the paramagnetic behavior dominates. Thus, despite its universal occurrence, diamagnetic behavior is observed only in a purely diamagnetic material. In a diamagnetic material, there are no unpaired electrons, so the intrinsic electron magnetic moments cannot produce any bulk effect. In these cases, the magnetization arises from the electrons' orbital motions, which can be understood classically as follows:

When a material is put in a magnetic field, the electrons circling the nucleus will experience, in addition to their Coulomb attraction to the nucleus, a Lorentz force from the magnetic field. Depending on which direction the electron is orbiting, this force may increase the centripetal force on the electrons, pulling them in towards the nucleus, or it may decrease the force, pulling them away from the nucleus. This effect systematically increases the orbital magnetic moments that were aligned opposite the field, and decreases the ones aligned parallel to the field (in accordance with Lenz›s law). This results in a small bulk magnetic moment, with an opposite direction to the applied field.

Note that this description is meant only as a heuristic; a proper understanding requires a quantum-mechanical description.

Note that all materials undergo this orbital response. However, in paramagnetic and ferromagnetic substances, the diamagnetic effect is overwhelmed by the much stronger effects caused by the unpaired electrons.

Paramagnetism

In a paramagnetic material there are *unpaired electrons*, i.e. atomic or molecular orbitals with exactly one electron in them. While paired electrons are required by the Pauli exclusion principle to have their intrinsic ('spin') magnetic moments pointing in opposite directions, causing their magnetic fields to cancel out, an unpaired electron is free to align its magnetic moment in any direction. When an external magnetic field is applied, these magnetic moments will tend to align themselves in the same direction as the applied field, thus reinforcing it.

Ferromagnetism

A ferromagnet, like a paramagnetic substance, has unpaired electrons. However, in *addition* to the electrons' intrinsic magnetic moment's tendency to be parallel to an *applied field*, there is also in these materials a tendency for these magnetic moments to orient parallel to *each other* to maintain a lowered-energy state. Thus, even in the absence of an applied field, the magnetic moments of the electrons in the material spontaneously line up parallel to one another.

Every ferromagnetic substance has its own individual temperature, called the Curie temperature, or Curie point, above which it loses its ferromagnetic properties. This is because the thermal tendency to disorder overwhelms the energy-lowering due to ferromagnetic order.

Tip of permanent magnet with coins demonstrating ferromagnetism

Ferromagnetism only occurs in a few substances; the common ones are iron, nickel, cobalt, their alloys, and some alloys of rare earth metals.

Magnetic Domains

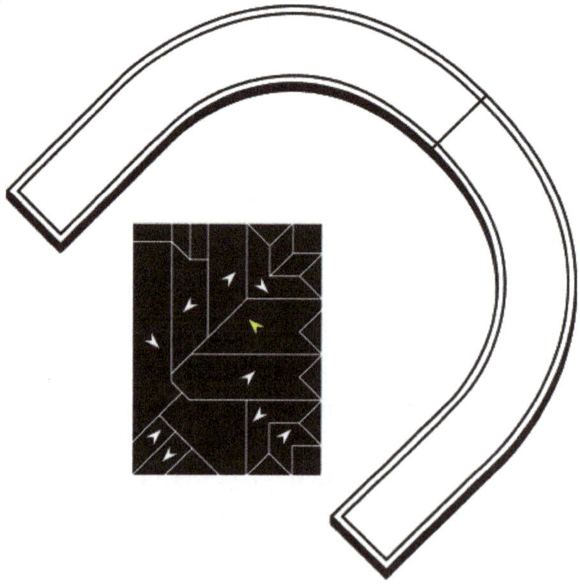

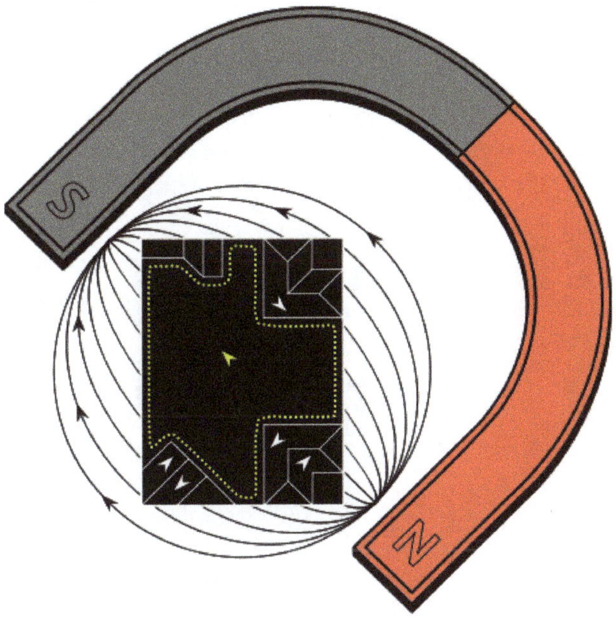

Left: Magnetic domains boundaries (white lines) in ferromagnetic material (black rectangle). **Right:** Effect of a magnet on the domains.

The magnetic moments of atoms in a ferromagnetic material cause them to behave something like tiny permanent magnets. They stick together and align themselves into small regions of more or less uniform alignment called magnetic domains or Weiss domains. Magnetic domains can be observed with a magnetic force microscope to reveal magnetic domain boundaries that resemble white lines in the sketch. There are many scientific experiments that can physically show magnetic fields.

When a domain contains too many molecules, it becomes unstable and divides into two domains aligned in opposite directions so that they stick together more stably as shown at the right.

When exposed to a magnetic field, the domain boundaries move so that the domains aligned with the magnetic field grow and dominate the structure (dotted yellow area) as shown at the left. When the magnetizing field is removed, the domains may not return to an unmagnetized state. This results in the ferromagnetic material›s being magnetized, forming a permanent magnet.

When magnetized strongly enough that the prevailing domain overruns all others to result in only one single domain, the material is magnetically saturated. When a magnetized ferromagnetic material is heated to the Curie point temperature, the molecules are agitated to the point that the magnetic domains lose the organization and the magnetic properties they cause cease. When the material is cooled, this domain alignment structure spontaneously returns, in a manner roughly analogous to how a liquid can freeze into a crystalline solid.

Antiferromagnetism

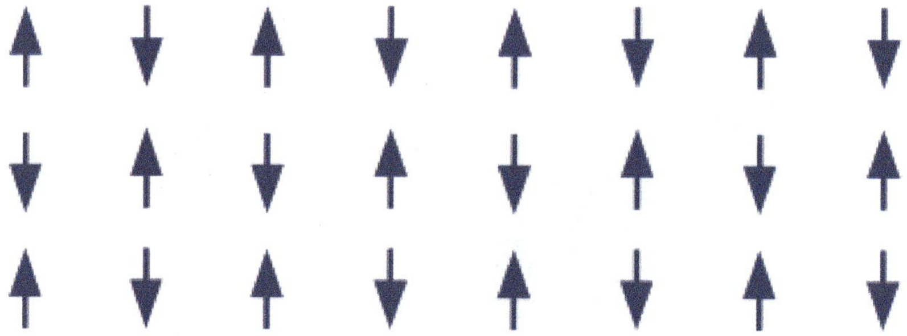

Antiferromagnetic ordering

In an antiferromagnet, unlike a ferromagnet, there is a tendency for the intrinsic magnetic moments of neighboring valence electrons to point in *opposite* directions. When all atoms are arranged in a substance so that each neighbor is 'anti-aligned', the substance is **antiferromagnetic**. Antiferromagnets have a zero net magnetic moment, meaning no field is produced by them. Antiferromagnets are less common compared to the other types of behaviors, and are mostly observed at low temperatures. In varying temperatures, antiferromagnets can be seen to exhibit diamagnetic and ferromagnetic properties.

In some materials, neighboring electrons want to point in opposite directions, but there is no geometrical arrangement in which *each* pair of neighbors is anti-aligned. This is called a spin glass, and is an example of geometrical frustration.

Ferrimagnetism

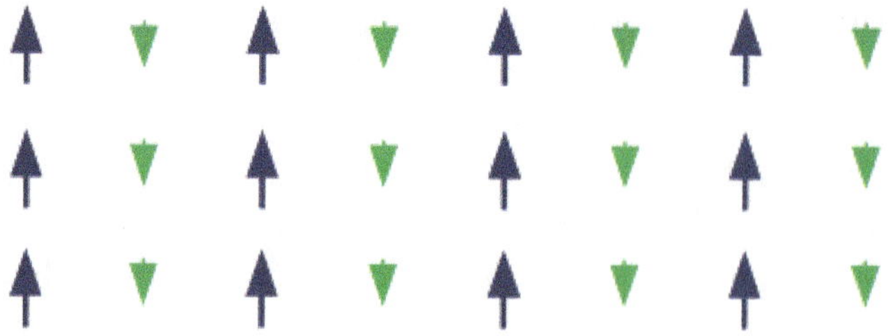

Ferrimagnetic ordering.

Like ferromagnetism, **ferrimagnets** retain their magnetization in the absence of a field. However, like antiferromagnets, neighboring pairs of electron spins tend to point in opposite directions. These two properties are not contradictory, because in the optimal geometrical arrangement, there is more magnetic moment from the sublattice of electrons that point in one direction, than from the sublattice that points in the opposite direction.

Most ferrites are ferrimagnetic. The first discovered magnetic substance, magnetite, is a ferrite and was originally believed to be a ferromagnet; Louis Néel disproved this, however, after discovering ferrimagnetism.

Superparamagnetism

When a ferromagnet or ferrimagnet is sufficiently small, it acts like a single magnetic spin that is subject to Brownian motion. Its response to a magnetic field is qualitatively similar to the response of a paramagnet, but much larger.

Other types of magnetism

- Metamagnetism
- Molecule-based magnet
- Spin glass

ELECTROMAGNET

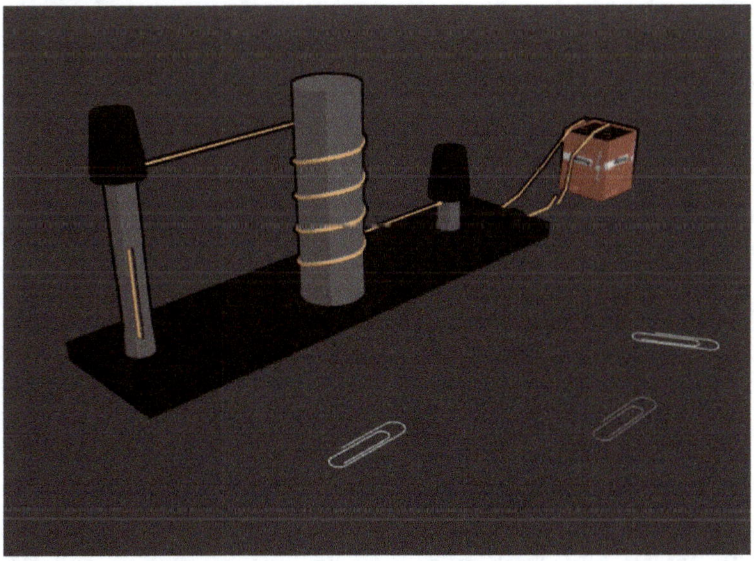

An electromagnet attracts paper clips when current is applied creating a magnetic field. The electromagnet loses them when current and magnetic field are removed.

An electromagnet is a type of magnet in which the magnetic field is produced by an electric current. The magnetic field disappears when the current is turned off. Electromagnets usually consist of a large number of closely spaced turns of wire that create the magnetic field. The wire turns are often wound around a magnetic core made from a ferromagnetic or ferrimagnetic material such as iron; the magnetic core concentrates the magnetic flux and makes a more powerful magnet.

The main advantage of an electromagnet over a permanent magnet is that the magnetic field can be quickly changed by controlling the amount of electric current in the winding. However, unlike a permanent magnet that needs no power, an electromagnet requires a continuous supply of current to maintain the magnetic field.

Electromagnets are widely used as components of other electrical devices, such as motors, generators, relays, loudspeakers, hard disks, MRI machines, scientific instruments, and magnetic separation equipment. Electromagnets are also employed in industry for picking up and moving heavy iron objects such as scrap iron and steel. Electromagnetism was discovered in 1820.

MAGNETISM, ELECTRICITY, AND SPECIAL RELATIVITY

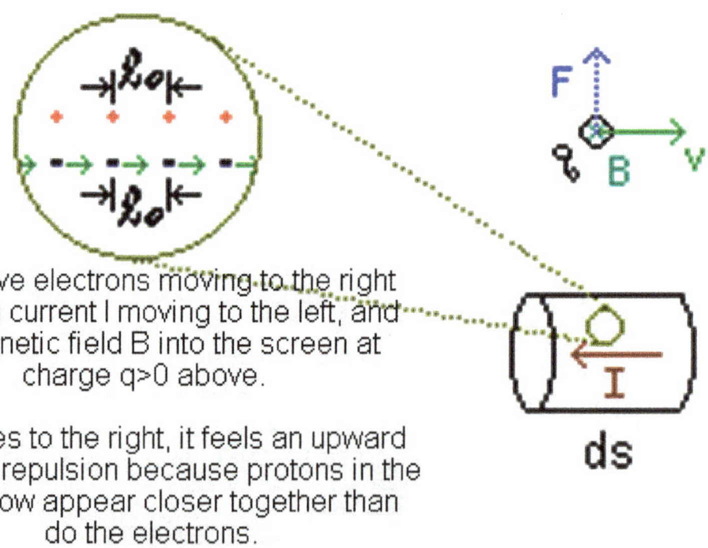

Magnetism from length-contraction.

As a consequence of Einstein's theory of special relativity, electricity and magnetism are fundamentally interlinked. Both magnetism lacking electricity, and electricity without magnetism, are inconsistent with special relativity, due to such effects as length contraction, time dilation, and the fact that the magnetic force is velocity-dependent. However, when both electricity and magnetism are taken into account, the resulting theory (electromagnetism) is fully consistent with special relativity. In particular, a phenomenon that appears purely electric or purely magnetic to one observer may be a mix of both to another, or more generally the relative contributions of electricity and magnetism are dependent on the frame of reference. Thus, special relativity mixes electricity and magnetism into a single, inseparable phenomenon called electromagnetism, analogous to how relativity mixes space and time into spacetime.

All observations on electromagnetism apply to what might be considered to be primarily magnetism, e.g. perturbations in the magnetic field are necessarily accompanied by a nonzero electric field, and propagate at the speed of light.

MAGNETIC FIELDS IN A MATERIAL

In a vacuum,

$$\mathbf{B} = \mu_0 \mathbf{H},$$

where μ_0 is the vacuum permeability.

In a material,

$$\mathbf{B} = \mu_0 (\mathbf{H} + \mathbf{M}).$$

The quantity $\mu_0\mathbf{M}$ is called *magnetic polarization.*

If the field **H** is small, the response of the magnetization **M** in a diamagnet or paramagnet is approximately linear:

$$\mathbf{M} = \chi \mathbf{H},$$

the constant of proportionality being called the magnetic susceptibility. If so,

$$\mu_0 (\mathbf{H} + \mathbf{M}) = \mu_0 (1 + \chi) \mathbf{H} = \mu_r \mu_0 \mathbf{H} = \mu \mathbf{H}.$$

In a hard magnet such as a ferromagnet, **M** is not proportional to the field and is generally nonzero even when **H** is zero (see Remanence).

Magnetic force

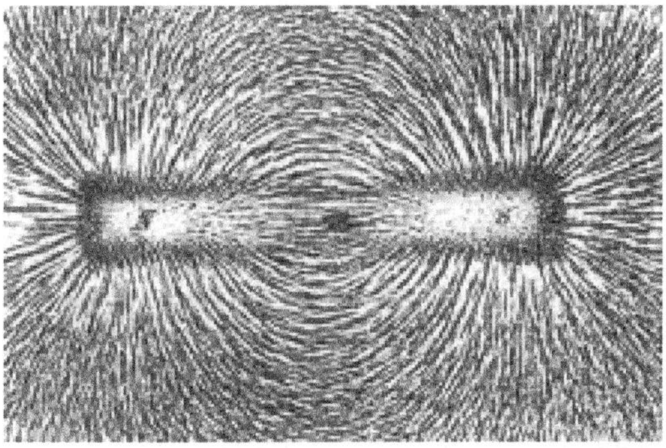

Magnetic lines of force of a bar magnet shown by iron filings on paper

The phenomenon of magnetism is "mediated" by the magnetic field. An electric current or magnetic dipole creates a magnetic field, and that field, in turn, imparts magnetic forces on other particles that are in the fields.

Maxwell's equations, which simplify to the Biot–Savart law in the case of steady currents, describe the origin and behavior of the fields that govern these forces. Therefore, magnetism is seen whenever electrically charged particles are in motion—for example, from movement of electrons in an electric current, or in certain cases from the orbital motion of electrons around an atom›s nucleus. They also arise from intrinsic magnetic dipoles arising from quantum-mechanical spin.

The same situations that create magnetic fields—charge moving in a current or in an atom, and intrinsic magnetic dipoles—are also the situations in which a magnetic field has an effect, creating a force. Following is the formula for moving charge; for the forces on an intrinsic dipole, see magnetic dipole.

When a charged particle moves through a magnetic field **B**, it feels a Lorentz force **F** given by the cross product:

$$\mathbf{F} = q(\mathbf{v} \times \mathbf{B})$$

where

q is the electric charge of the particle, and

v is the velocity vector of the particle

Because this is a cross product, the force is perpendicular to both the motion of the particle and the magnetic field. It follows that the magnetic force does no work on the particle; it may change the direction of the particle's movement, but it cannot cause it to speed up or slow down. The magnitude of the force is

$$F = qvB \sin \theta$$

where θ is the angle between **v** and **B**.

One tool for determining the direction of the velocity vector of a moving charge, the magnetic field, and the force exerted is labeling the index finger "V", the middle finger "B", and the thumb "F" with your right hand. When making a gun-like configuration, with the middle finger crossing under the index finger, the fingers represent the velocity vector, magnetic field vector, and force vector, respectively. See also right hand rule.

Magnetic Dipoles

A very common source of magnetic field shown in nature is a dipole, with a "South pole" and a "North pole", terms dating back to the use of magnets as compasses, interacting with the Earth's magnetic field to indicate North and South on the globe. Since opposite ends of magnets are attracted, the north pole of a magnet is attracted to the south pole of another magnet. The Earth›s North Magnetic Pole (currently in the Arctic Ocean, north of Canada) is physically a south pole, as it attracts the north pole of a compass. A magnetic field contains energy, and physical systems move toward configurations with lower energy. When diamagnetic material is placed in a magnetic field, a *magnetic dipole* tends to align itself in opposed polarity to that field, thereby lowering the net field strength. When ferromagnetic material is placed within a magnetic field, the magnetic dipoles align to the applied field, thus expanding the domain walls of the magnetic domains.

Magnetic Monopoles

Since a bar magnet gets its ferromagnetism from electrons distributed evenly throughout the bar, when a bar magnet is cut in half, each of the resulting pieces is a smaller bar magnet. Even though a magnet is said to have a north pole and a south pole, these two poles cannot be separated from each other. A monopole—if such a thing exists—would be a new and fundamentally different kind of magnetic object. It would act as an isolated north pole, not attached to a south pole, or vice versa. Monopoles would carry "magnetic charge" analogous to electric charge. Despite systematic searches since 1931, as of 2010, they have never been observed, and could very well not exist.

Nevertheless, some theoretical physics models predict the existence of these magnetic monopoles. Paul Dirac observed in 1931 that, because electricity and magnetism show a certain symmetry, just as quantum theory predicts that individual positive or negative electric charges can be observed without the opposing charge, isolated South or North magnetic poles should be observable. Using quantum theory Dirac showed that if magnetic monopoles exist, then one could explain the quantization of electric charge—that is, why the observed elementary particles carry charges that are multiples of the charge of the electron.

Certain grand unified theories predict the existence of monopoles which, unlike elementary particles, are solitons (localized energy packets). The initial results of using these models to estimate the number of monopoles created in the big bang contradicted cosmological observations—the monopoles would have been so plentiful and massive that they would have long since halted the expansion of the universe. However, the idea of inflation (for which this problem served as a partial motivation) was successful in solving this problem, creating models in which monopoles existed but were rare enough to be consistent with current observations.

QUANTUM-MECHANICAL ORIGIN OF MAGNETISM

In principle all kinds of magnetism originate (similar to superconductivity) from specific quantum-mechanical phenomena (e.g. Mathematical formulation of quantum mechanics, in particular the chapters on spin and on the Pauli principle). A successful model was developed already in 1927, by Walter Heitler and Fritz London, who derived quantum-mechanically, how hydrogen molecules are formed from hydrogen atoms, i.e. from the atomic hydrogen orbitals u_A and u_B centered at the nuclei *A* and *B*, see below. That this leads to magnetism is not at all obvious, but will be explained in the following.

According to the Heitler-London theory, so-called two-body molecular σ-orbitals are formed, namely the resulting orbital is:

$$\psi(\mathbf{r}_1, \mathbf{r}_2) = \frac{1}{\sqrt{2}} \left(u_A(\mathbf{r}_1)u_B(\mathbf{r}_2) + u_B(\mathbf{r}_1)u_A(\mathbf{r}_2)\right)$$

Here the last product means that a first electron, $\mathbf{r}_1$, is in an atomic hydrogen-orbital centered at the second nucleus, whereas the second electron runs around the first nucleus. This "exchange" phenomenon is an expression for the quantum-mechanical property that particles with identical properties cannot be distinguished. It is specific not only for the formation of chemical bonds, but as we will see, also for magnetism, i.e. in this connection the term exchange interaction arises, a term which is essential for the origin of magnetism, and which is stronger, roughly by factors 100 and even by 1000, than the energies arising from the electrodynamic dipole-dipole interaction.

As for the *spin function* $\chi(s_1, s_2)$, which is responsible for the magnetism, we have the already mentioned Pauli's principle, namely that a symmetric orbital (i.e. with the + sign as above) must be multiplied with an antisymmetric spin function (i.e. with a − sign), and *vice versa*. Thus:

$$\chi(s_1, s_2) = \frac{1}{\sqrt{2}} (\alpha(s_1)\beta(s_2) - \beta(s_1)\alpha(s_2)),$$

I.e., not only u_A and u_B must be substituted by α and β, respectively (the first entity means spin up, the second one spin down), but also the sign + by the − sign, and finally $\mathbf{r}_i$ by the discrete values s_i (= ±½); thereby we have $\alpha(+1/2) = \beta(-1/2) = 1$ and $\alpha(-1/2) = \beta(+1/2) = 0$. The "singlet state", i.e. the − sign, means: the spins are *antiparallel*, i.e. for the solid we have antiferromagnetism, and for two-atomic molecules one has diamagnetism. The tendency to form a (homoeopolar) chemical bond (this means: the formation of a *symmetric* molecular orbital, i.e. with the + sign) results through the Pauli principle automatically in an *antisymmetric* spin state (i.e. with the − sign). In contrast, the Coulomb repulsion of the electrons, i.e. the tendency that they try to avoid each other by this repulsion, would lead to an *antisymmetric* orbital function (i.e. with the − sign) of these two particles, and complementary to a *symmetric* spin function (i.e. with the + sign, one of the so-called "triplet functions). Thus, now the spins would be *parallel* (ferromagnetism in a solid, paramagnetism in two-atomic gases).

The last-mentioned tendency dominates in the metals iron, cobalt and nickel, and in some rare earths, which are *ferromagnetic*. Most of the other metals, where the first-mentioned tendency dominates, are *nonmagnetic* (e.g. sodium, aluminium, and magnesium) or *antiferromagnetic* (e.g. manganese). Diatomic gases are also almost exclusively diamagnetic, and not paramagnetic. However, the oxygen molecule, because of the involvement of π-orbitals, is an exception important for the life-sciences.

The Heitler-London considerations can be generalized to the Heisenberg model of magnetism (Heisenberg 1928).

The explanation of the phenomena is thus essentially based on all subtleties of quantum mechanics, whereas the electrodynamics covers mainly the phenomenology.

UNITS

SI

SI electromagnetism units			
Symbol[17]	Name of quantity	Derived units	Conversion of international to SI base units
I	Electric current	ampere (SI base unit)	$\mathbf{A = C\,s^{-1}}$
q	Electric charge	coulomb	$\mathbf{C = A\,s}$
U, ΔV, $\Delta\phi$, **E**	Potential difference; Electromotive force	volt	$\mathbf{V = J\,C^{-1} = kg\,A^{-1}m^2s^{-3}}$
R; Z; X	Resistance; Impedance; Reactance	ohm	$\mathbf{\Omega = V\,A^{-1} = kg\,m^2\,A^{-2}s^{-3}}$
ρ	Resistivity	ohm metre	$\mathbf{\Omega\,m = kg\,A^{-2}m^3s^{-3}}$
P	Electric power	watt	$\mathbf{W = V\,A = kg\,m^2s^{-3}}$
C	Capacitance	farad	$\mathbf{F = C\,V^{-1} = A^2kg^{-1}m^{-2}s^4}$
E	Electric field strength	volt per metre	$\mathbf{V\,m^{-1} = C^{-1}N = kg\,A^{-1}m\,s^{-3}}$
D	Electric displacement field	Coulomb per square metre	$\mathbf{C\,m^{-2} = A\,m^{-2}s}$
ε	Permittivity	farad per metre	$\mathbf{F\,m^{-1} = A^2kg^{-1}m^{-3}s^4}$
χ_e	Electric susceptibility	Dimensionless	
G; Υ; **B**	Conductance; Admittance; Susceptance	siemens	$\mathbf{S = \Omega^{-1} = kg^{-1}A^2m^{-2}s^3}$
γ, κ, σ	Conductivity	siemens per metre	$\mathbf{S\,m^{-1} = A^2kg^{-1}m^{-3}s^3}$
B	Magnetic flux density, Magnetic induction	tesla	$\mathbf{T = Wb\,m^{-2} = kg\,A^{-1}s^{-2}}$
Φ	Magnetic flux	weber	$\mathbf{Wb = V\,s = kg\,A^{-1}m^2s^{-2}}$
H	Magnetic field strength	ampere per metre	$\mathbf{A\,m^{-1}}$
L, **M**	Inductance	henry	$\mathbf{H = Wb\,A^{-1} = V\,A^{-1}s = kg\,A^{-2}m^2s^{-2}}$
μ	Permeability	henry per metre	$\mathbf{Hm^{-1} = kg\,A^{-2}m\,s^{-2}}$
χ	Magnetic susceptibility	Dimensionless	

Other

- gauss – the centimeter-gram-second (CGS) unit of magnetic field (denoted **B**)
- oersted – the CGS unit of magnetizing field (denoted **H**)
- maxwell – the CGS unit for magnetic flux
- gamma – a unit of *magnetic flux density* that was commonly used before the tesla came into use (1.0 gamma = 1.0 nanotesla)

- μ_0 – common symbol for the permeability of free space ($4\pi\times10^{-7}$ newton/(ampere-turn)^2)

LIVING THINGS

Some organisms can detect magnetic fields, a phenomenon known as magnetoception. In addition to detection, biomagnetic pheonomena is utilized by organisms in a number of ways. For instance, chitons, a type of marine mollusk, produce magnetite to harden their teeth, and even humans produce magnetite in bodily tissue. Magnetobiology studies magnetic fields as a medical treatment; fields naturally produced by an organism are known as biomagnetism.

REFERENCES

1. Du Trémolet de Lacheisserie, Étienne; Damien Gignoux; Michel Schlenker (2005). Magnetism: Fundamentals. Springer. pp. 3–6. ISBN 0-387-22967-1.
2. Fowler, Michael (1997). Historical Beginnings of Theories of Electricity and Magnetism. Retrieved 2008-04-02.
3. Vowles, Hugh P. (1932). Early Evolution of Power Engineering. Isis. University of Chicago Press. 17 (2): 412–420 [419–20]. doi:10.1086/346662.
4. Li, Shu-hua (1954). Origine de la Boussole II. Aimant et Boussole. Isis (in French). 45 (2): 175–196. JSTOR 227361. un passage dans le Liu-che-tch›ouen-ts›ieou [...]: "La pierre d›aimant fait venir le fer ou elle l›attire."
5. Schmidl, Petra G. (1996–1997). Two Early Arabic Sources On The Magnetic Compass. Journal of Arabic and Islamic Studies. 1: 81–132.
6. A. Einstein: On the Electrodynamics of Moving Bodies, June 30, 1905.
7. HP Meyers (1997). Introductory solid state physics (2 ed.). CRC Press. p. 362; Figure 11.1. ISBN 9781420075021.
8. Catherine Westbrook; Carolyn Kaut; Carolyn Kaut-Roth (1998). MRI (Magnetic Resonance Imaging) in practice (2 ed.). Wiley-Blackwell. p. 217. ISBN 0-632-04205-2.
9. Merzouki, Rochdi; Samantaray, Arun Kumar; Pathak, Pushparaj Mani (2012). Intelligent Mechatronic Systems: Modeling, Control and Diagnosis. Springer Science & Business Media. pp. 403–405. ISBN 144714628X.
10. Sturgeon, W. (1825). Improved Electro Magnetic Apparatus. Trans. Royal Society of Arts, Manufactures, & Commerce. London. 43: 37–52. cited in Miller, T.J.E (2001). Electronic Control of Switched Reluctance Machines. Newnes. p. 7. ISBN 0-7506-5073-7.
11. Griffiths 1998, chapter 12
12. Jackson, John David (1999). Classical electrodynamics (3rd ed.). New York: Wiley. ISBN 0-471-30932-X.

13. Milton mentions some inconclusive events (p.60) and still concludes that "no evidence at all of magnetic monopoles has survived" (p.3). Milton, Kimball A. (June 2006). Theoretical and experimental status of magnetic monopoles. Reports on Progress in Physics. 69 (6): 1637–1711. arXiv:hep-ex/0602040. Bibcode:2006RPPh...69.1637M. doi:10.1088/0034-4885/69/6/R02..

14. Guth, Alan (1997). The Inflationary Universe: The Quest for a New Theory of Cosmic Origins. Perseus. ISBN 0-201-32840-2. OCLC 38941224..

15. International Union of Pure and Applied Chemistry (1993). Quantities, Units and Symbols in Physical Chemistry, 2nd edition, Oxford: Blackwell Science. ISBN 0-632-03583-8. pp. 14–15. Electronic version.

16. Kirschvink, Joseph L.; Kobayashi-Kirshvink, Atsuko; Diaz-Ricci, Juan C.; Kirschvink, Steven J. (1992). Magnetite in Human Tissues: A Mechanism for the Biological Effects of Weak ELF Magnetic Fields (PDF). Bioelectromagnetics Supplement. 1: 101–113. Retrieved 29 March 2016.

CHAPTER 6

Superconductivity

Chapter Contents

Superconductivity is a property displayed by certain materials at very low temperatures. Materials found to have this property include metals and their alloys (tin, aluminum, and others), some semiconductors, and certain ceramics known as cuprates that contain copper and oxygen atoms. A superconductor conducts electricity without resistance, a unique property. It also repels magnetic fields perfectly in a phenomenon known as the Meissner effect, losing any internal magnetic field it might have had before being cooled to a critical temperature. Because of this effect, some can be made to float endlessly above a strong magnetic field.

INTRODUCTION

Superconductivity is a phenomenon of exactly zero electrical resistance and expulsion of magnetic flux fields occurring in certain materials, called superconductors, when cooled below a characteristic critical temperature. It was discovered by Dutch physicist Heike Kamerlingh Onnes on April 8, 1911, in Leiden. Like ferromagnetism and atomic spectral lines, superconductivity is a quantum mechanical phenomenon. It is characterized by the Meissner effect, the complete ejection of magnetic field lines from the interior of the superconductor as it transitions into the superconducting state. The occurrence of the Meissner effect indicates that superconductivity cannot be understood simply as the idealization of *perfect conductivity* in classical physics.

The electrical resistance of a metallic conductor decreases gradually as temperature is lowered. In ordinary conductors, such as copper or silver, this decrease is limited by impurities and other defects. Even near absolute zero, a real sample of a normal conductor shows some resistance. In a superconductor, the resistance drops abruptly to zero when the material is cooled below its critical temperature. An electric current flowing through a loop of superconducting wire can persist indefinitely with no power source.

A magnet levitating above a high-temperature superconductor, cooled with liquid nitrogen. Persistent electric current flows on the surface of the superconductor, acting to exclude the magnetic field of the magnet (Faraday›s law of induction). This current effectively forms an electromagnet that repels the magnet.

A high-temperature superconductor levitating above a magnet.

In 1986, it was discovered that some cuprate-perovskite ceramic materials have a critical temperature above 90 K (−183 °C). Such a high transition temperature is theoretically impossible for a conventional superconductor, leading the materials to be termed high-temperature superconductors. The cheaply-available coolant liquid nitrogen boils at 77 K, and thus superconduction at higher temperatures than this facilitates many experiments and applications that are less practical at lower temperatures.

CLASSIFICATION

There are many criteria by which superconductors are classified. The most common are:

- **Response to a magnetic field**: A superconductor can be *Type I*, meaning it has a single critical field, above which all superconductivity is lost; or *Type II*, meaning it has two critical fields, between which it allows partial penetration of the magnetic field.
- **By theory of operation**: It is *conventional* if it can be explained by the BCS theory or its derivatives, or *unconventional*, otherwise.
- **By critical temperature**: A superconductor is generally considered *high-temperature* if it reaches a superconducting state when cooled using liquid nitrogen – that is, at only $T_c > 77$ K) – or *low-temperature* if more aggressive cooling techniques are required to reach its critical temperature.
- **By material**: Superconductor material classes include chemical elements (e.g. mercury or lead), alloys (such as niobium-titanium, germanium-niobium, and niobium nitride), ceramics (YBCO and magnesium diboride), superconducting pnictides (like fluorine-doped LaOFeAs) or organic superconductors (fullerenes and carbon nanotubes; though perhaps these examples should be included among the chemical elements, as they are composed entirely of carbon).

Elementary Properties of Superconductors

Most of the physical properties of superconductors vary from material to material, such as the heat capacity and the critical temperature, critical field, and critical current density at which superconductivity is destroyed.

On the other hand, there is a class of properties that are independent of the underlying material. For instance, all superconductors have *exactly* zero resistivity to low applied currents when there is no magnetic field present or if the applied field does not exceed a critical value. The existence of these "universal" properties implies that superconductivity is a thermodynamic phase, and thus possesses certain distinguishing properties which are largely independent of microscopic details.

Zero Electrical DC Resistance

Electric cables for accelerators at CERN. Both the massive and slim cables are rated for 12,500 A. *Top:* conventional cables for LEP; *bottom:* superconductor-based cables for the LHC.

Cross section of a preform superconductor rod from abandoned Texas Superconducting Super Collider (SSC).

The simplest method to measure the electrical resistance of a sample of some material is to place it in an electrical circuit in series with a current source I and measure the resulting voltage V across the sample. The resistance of the sample is given by Ohm›s law as $R = V / I$. If the voltage is zero, this means that the resistance is zero.

Superconductors are also able to maintain a current with no applied voltage whatsoever, a property exploited in superconducting electromagnets such as those found in MRI machines. Experiments have demonstrated that currents in superconducting coils can persist for years without any measurable degradation. Experimental evidence points to a current lifetime of at least 100,000 years. Theoretical estimates for the lifetime of a persistent current can exceed the estimated lifetime of the universe, depending on the wire geometry and the temperature.

In a normal conductor, an electric current may be visualized as a fluid of electrons moving across a heavy ionic lattice. The electrons are constantly colliding with the ions in the lattice, and during each collision some of the energy carried by the current is absorbed by the lattice and converted into heat, which is essentially the vibrational kinetic energy of the lattice ions. As a result, the energy carried by the current is constantly being dissipated. This is the phenomenon of electrical resistance and Joule heating.

The situation is different in a superconductor. In a conventional superconductor, the electronic fluid cannot be resolved into individual electrons. Instead, it consists of bound *pairs* of electrons known as Cooper pairs. This pairing is caused by an attractive force between electrons from the exchange of phonons. Due to quantum mechanics, the energy spectrum of this Cooper pair fluid possesses an *energy gap*, meaning there is a minimum amount of energy ΔE that must be supplied in order to excite the fluid. Therefore, if ΔE is larger than the thermal energy of the lattice, given by kT, where k is Boltzmann›s constant and T is the temperature, the fluid will not be scattered by the lattice. The Cooper pair fluid is thus a superfluid, meaning it can flow without energy dissipation.

In a class of superconductors known as type II superconductors, including all known high-temperature superconductors, an extremely small amount of resistivity appears at temperatures not too far below the nominal superconducting transition when an electric current is applied in conjunction with a strong magnetic field, which may be caused by the electric current. This is due to the motion of magnetic vortices in the electronic superfluid, which dissipates some of the energy carried by the current. If the current is sufficiently small, the vortices are stationary, and the resistivity vanishes. The resistance due to this effect is tiny compared with that of non-superconducting materials, but must be taken into account in sensitive experiments. However, as the temperature decreases far enough below the nominal superconducting transition, these vortices can become frozen into a disordered but stationary phase known as a "vortex glass". Below this vortex glass transition temperature, the resistance of the material becomes truly zero.

Superconducting Phase Transition

In superconducting materials, the characteristics of superconductivity appear when the temperature T is lowered below a **critical temperature** T_c. The value of this critical temperature varies from material to material. Conventional superconductors usually have critical temperatures ranging from around 20 K to less than 1 K. Solid mercury, for example, has a critical temperature of 4.2 K. As of 2009, the highest critical temperature found for a conventional superconductor is 39 K for magnesium diboride (MgB_2), although this material displays enough exotic properties that there is some doubt about classifying it as a "conventional" superconductor. Cuprate superconductors can have much higher critical temperatures: $YBa_2Cu_3O_7$, one of the first cuprate superconductors to be discovered, has a critical temperature of 92 K, and mercury-based cuprates have been found with critical temperatures in excess of 130 K. The explanation for these high critical temperatures remains unknown. Electron pairing due to phonon exchanges explains superconductivity in conventional superconductors, but it does not explain superconductivity in the newer superconductors that have a very high critical temperature.

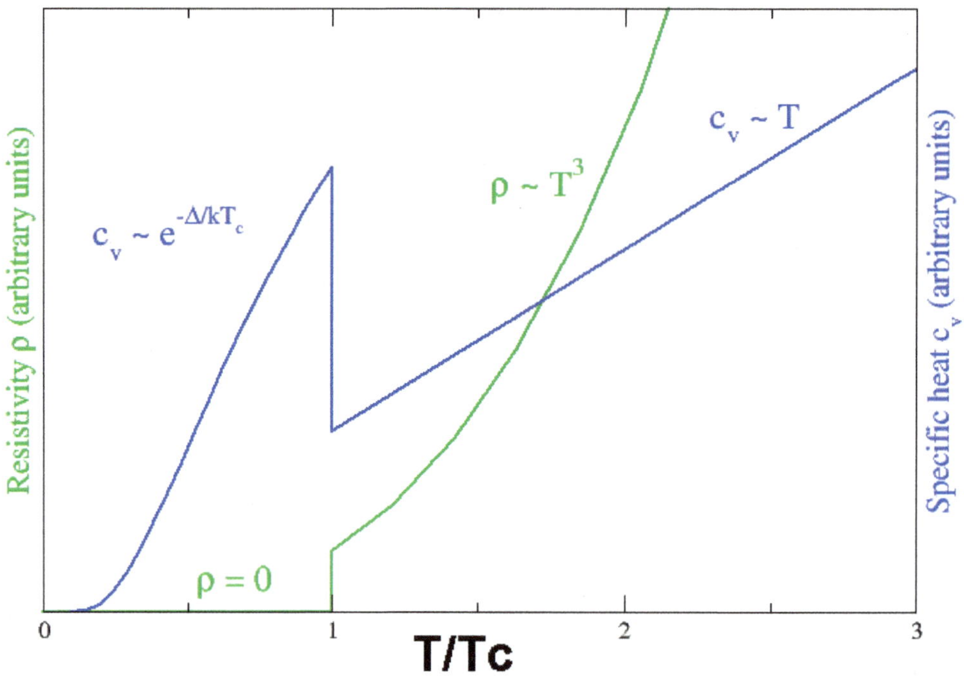

Behavior of heat capacity (c_v, blue) and resistivity (ρ, green) at the superconducting phase transition.

Similarly, at a fixed temperature below the critical temperature, superconducting materials cease to superconduct when an external magnetic field is applied which is greater than the *critical magnetic field*. This is because the Gibbs free energy of the superconducting phase increases quadratically with the magnetic field while the free energy of the normal phase is roughly independent of the magnetic field. If the material superconducts in the absence of a field, then the superconducting phase free energy is lower than that of the normal phase and so for some finite value of the magnetic field (proportional to the square root of the difference of the free energies at zero magnetic field) the two free energies will be equal and a phase transition to the normal phase will occur. More generally, a higher temperature and a stronger magnetic field lead to a smaller fraction of the electrons in the superconducting band and consequently a longer London penetration depth of external magnetic fields and currents. The penetration depth becomes infinite at the phase transition.

The onset of superconductivity is accompanied by abrupt changes in various physical properties, which is the hallmark of a phase transition. For example, the electronic heat capacity is proportional to the temperature in the normal (non-superconducting) regime. At the superconducting transition, it suffers a discontinuous jump and thereafter ceases to be linear. At low temperatures, it varies instead as $e^{-\alpha/T}$ for some constant, α. This exponential behavior is one of the pieces of evidence for the existence of the energy gap.

The order of the superconducting phase transition was long a matter of debate. Experiments indicate that the transition is second-order, meaning there is no latent heat. However, in the presence of an external magnetic field there is latent heat, because the superconducting phase has a lower entropy below the critical temperature than the normal phase. It has been experimentally demonstrated that, as a consequence, when the magnetic field is increased beyond the critical field, the resulting phase transition leads to a decrease in the temperature of the superconducting material.

Calculations in the 1970s suggested that it may actually be weakly first-order due to the effect of long-range fluctuations in the electromagnetic field. In the 1980s it was shown theoretically with the help of a disorder field theory, in which the vortex lines of the superconductor play a major role, that the transition is of second order within the type II regime and of first order (i.e., latent heat) within the type I regime, and that the two regions are separated by a tricritical point. The results were strongly supported by Monte Carlo computer simulations.

Meissner Effect

When a superconductor is placed in a weak external magnetic field **H**, and cooled below its transition temperature, the magnetic field is ejected. The Meissner effect does not cause the field to be completely ejected but instead the field penetrates the superconductor but only to a very small distance, characterized by a parameter λ, called the London penetration depth, decaying exponentially to zero within the bulk of the material. The Meissner effect is a defining characteristic of superconductivity. For most superconductors, the London penetration depth is on the order of 100 nm.

The Meissner effect is sometimes confused with the kind of diamagnetism one would expect in a perfect electrical conductor: according to Lenz›s law, when a *changing* magnetic field is applied to a conductor, it will induce an electric current in the conductor that creates an opposing magnetic field. In a perfect conductor, an arbitrarily large current can be induced, and the resulting magnetic field exactly cancels the applied field.

The Meissner effect is distinct from this—it is the spontaneous expulsion which occurs during transition to superconductivity. Suppose we have a material in its normal state, containing a constant internal magnetic field. When the material is cooled below the critical temperature, we would observe the abrupt expulsion of the internal magnetic field, which we would not expect based on Lenz›s law.

The Meissner effect was given a phenomenological explanation by the brothers Fritz and Heinz London, who showed that the electromagnetic free energy in a superconductor is minimized provided

$$\nabla^2 \mathbf{H} = \lambda^{-2} \mathbf{H}$$

where **H** is the magnetic field and λ is the London penetration depth.

This equation, which is known as the London equation, predicts that the magnetic field in a superconductor decays exponentially from whatever value it possesses at the surface.

A superconductor with little or no magnetic field within it is said to be in the Meissner state. The Meissner state breaks down when the applied magnetic field is too large. Superconductors can be divided into two classes according to how this breakdown occurs. In Type I superconductors, superconductivity is abruptly destroyed when the strength of the applied field rises above a critical value H_c. Depending on the geometry of the sample, one may obtain an intermediate state consisting of a baroque pattern of regions of normal material carrying a magnetic field mixed with regions of superconducting material containing no field. In Type II superconductors, raising the applied field past a critical value H_{c1} leads to a mixed state (also known as the vortex state) in which an increasing amount of magnetic flux penetrates the material, but there remains no resistance to the flow of electric current as long as the current is not too large. At a second critical field strength H_{c2}, superconductivity is destroyed. The mixed state is actually caused by vortices in the electronic superfluid, sometimes called fluxons because the flux carried by these vortices is quantized. Most pure elemental superconductors, except niobium and carbon nanotubes, are Type I, while almost all impure and compound superconductors are Type II.

London Moment

Conversely, a spinning superconductor generates a magnetic field, precisely aligned with the spin axis. The effect, the London moment, was put to good use in Gravity Probe B. This experiment measured the magnetic fields of four superconducting gyroscopes to determine their spin axes. This was critical to the experiment since it is one of the few ways to accurately determine the spin axis of an otherwise featureless sphere.

HISTORY OF SUPERCONDUCTIVITY

Heike Kamerlingh Onnes (right), the discoverer of superconductivity. Paul Ehrenfest, Hendrik Lorentz, Niels Bohr stand to his left.

Superconductivity was discovered on April 8, 1911 by Heike Kamerlingh Onnes, who was studying the resistance of solid mercury at cryogenic temperatures using the recently produced liquid helium as a refrigerant. At the temperature of 4.2 K, he observed that the resistance abruptly disappeared. In the same experiment, he also observed the superfluid transition of helium at 2.2 K, without recognizing its significance. The precise date and circumstances of the discovery were only reconstructed a century later, when Onnes's notebook was found. In subsequent decades, superconductivity was observed in several other materials. In 1913, lead was found to superconduct at 7 K, and in 1941 niobium nitride was found to superconduct at 16 K.

Great efforts have been devoted to finding out how and why superconductivity works; the important step occurred in 1933, when Meissner and Ochsenfeld discovered that superconductors expelled applied magnetic fields, a phenomenon which has come to be known as the Meissner effect. In 1935, Fritz and Heinz London showed that the Meissner effect was a consequence of the minimization of the electromagnetic free energy carried by superconducting current.

London Theory

The first phenomenological theory of superconductivity was London theory. It was put forward by the brothers Fritz and Heinz London in 1935, shortly after the discovery that magnetic fields are expelled from superconductors. A major triumph of the equations of this theory is their ability to explain the Meissner effect, wherein a material exponentially expels all internal magnetic fields as it crosses the superconducting threshold. By using the London equation, one can obtain the dependence of the magnetic field inside the superconductor on the distance to the surface.

There are two London equations:

$$\frac{\partial \mathbf{j}_s}{\partial t} = \frac{n_s e^2}{m}\mathbf{E}, \qquad \nabla \times \mathbf{j}_s = -\frac{n_s e^2}{m}\mathbf{B}.$$

The first equation follows from Newton's second law for superconducting electrons.

Conventional Theories (1950s)

During the 1950s, theoretical condensed matter physicists arrived at an understanding of "conventional" superconductivity, through a pair of remarkable and important theories: the phenomenological Ginzburg-Landau theory (1950) and the microscopic BCS theory (1957).

In 1950, the phenomenological Ginzburg-Landau theory of superconductivity was devised by Landau and Ginzburg. This theory, which combined Landau›s theory of second-order phase transitions with a Schrödinger-like wave equation, had great success in explaining the macroscopic properties of superconductors. In particular, Abrikosov showed that Ginzburg-Landau theory predicts the division of superconductors into the two categories now referred to as Type I and Type II. Abrikosov and Ginzburg were awarded the 2003 Nobel Prize for their work (Landau had received the 1962 Nobel Prize for other work, and died in 1968). The four-dimensional extension of the Ginzburg-Landau theory, the Coleman-Weinberg model, is important in quantum field theory and cosmology.

Also in 1950, Maxwell and Reynolds *et al.* found that the critical temperature of a superconductor depends on the isotopic mass of the constituent element. This important discovery pointed to the electron-phonon interaction as the microscopic mechanism responsible for superconductivity.

The complete microscopic theory of superconductivity was finally proposed in 1957 by Bardeen, Cooper and Schrieffer. This BCS theory explained the superconducting current as a superfluid of Cooper pairs, pairs of electrons interacting through the exchange of phonons. For this work, the authors were awarded the Nobel Prize in 1972.

The BCS theory was set on a firmer footing in 1958, when N. N. Bogolyubov showed that the BCS wavefunction, which had originally been derived from a variational argument, could be obtained using a canonical transformation of the electronic Hamiltonian. In 1959, Lev Gor›kov showed that the BCS theory reduced to the Ginzburg-Landau theory close to the critical temperature.

Generalizations of BCS theory for conventional superconductors form the basis for understanding of the phenomenon of superfluidity, because they fall into the lambda transition universality class. The extent to which such generalizations can be applied to unconventional superconductors is still controversial.

Further History

The first practical application of superconductivity was developed in 1954 with Dudley Allen Buck's invention of the cryotron. Two superconductors with greatly different values of critical magnetic field are combined to produce a fast, simple, switch for computer elements.

Soon after discovering superconductivity in 1911, Kamerlingh Onnes attempted to make an electromagnet with superconducting windings but found that relatively low magnetic fields destroyed superconductivity in the materials he investigated. Much later, in 1955, G.B. Yntema succeeded in constructing a small 0.7-tesla iron-core electromagnet with superconducting niobium wire windings. Then, in 1961, J.E. Kunzler, E. Buehler, F.S.L. Hsu, and J.H. Wernick made the startling discovery that, at 4.2 kelvin, a compound consisting of three parts niobium and one part tin, was capable of supporting a current density of more than 100,000 amperes per square centimeter in a magnetic field of 8.8 tesla. Despite being brittle and difficult to fabricate, niobium-tin has since proved extremely useful in supermagnets generating magnetic fields as high as 20 tesla. In 1962 T.G. Berlincourt and R.R. Hake discovered that alloys of niobium and titanium are suitable for applications up to 10 tesla. Promptly thereafter, commercial production of niobium-titanium supermagnet wire commenced at Westinghouse Electric Corporation and at Wah Chang Corporation. Although niobium-titanium boasts less-impressive superconducting properties than those of niobium-tin, niobium-titanium has, nevertheless, become the most widely used "workhorse" supermagnet material, in large measure a consequence of its very-high ductility and ease of fabrication. However, both niobium-tin and niobium-titanium find wide application in MRI medical imagers, bending and focusing magnets for enormous high-energy-particle accelerators, and a host of other applications. Conectus, a European superconductivity consortium, estimated that in 2014, global economic activity for which superconductivity was indispensable amounted to about five billion euros, with MRI systems accounting for about 80% of that total.

In 1962, Josephson made the important theoretical prediction that a supercurrent can flow between two pieces of superconductor separated by a thin layer of insulator. This phenomenon, now called the Josephson effect, is exploited by superconducting devices such as SQUIDs. It is used in the most accurate available measurements of the magnetic flux quantum $\Phi_0 = h/(2e)$, where h is the Planck constant. Coupled with the quantum Hall resistivity, this leads to a precise measurement of the Planck constant. Josephson was awarded the Nobel Prize for this work in 1973.

In 2008, it was proposed that the same mechanism that produces superconductivity could produce a superinsulator state in some materials, with almost infinite electrical resistance.

HIGH-TEMPERATURE SUPERCONDUCTIVITY

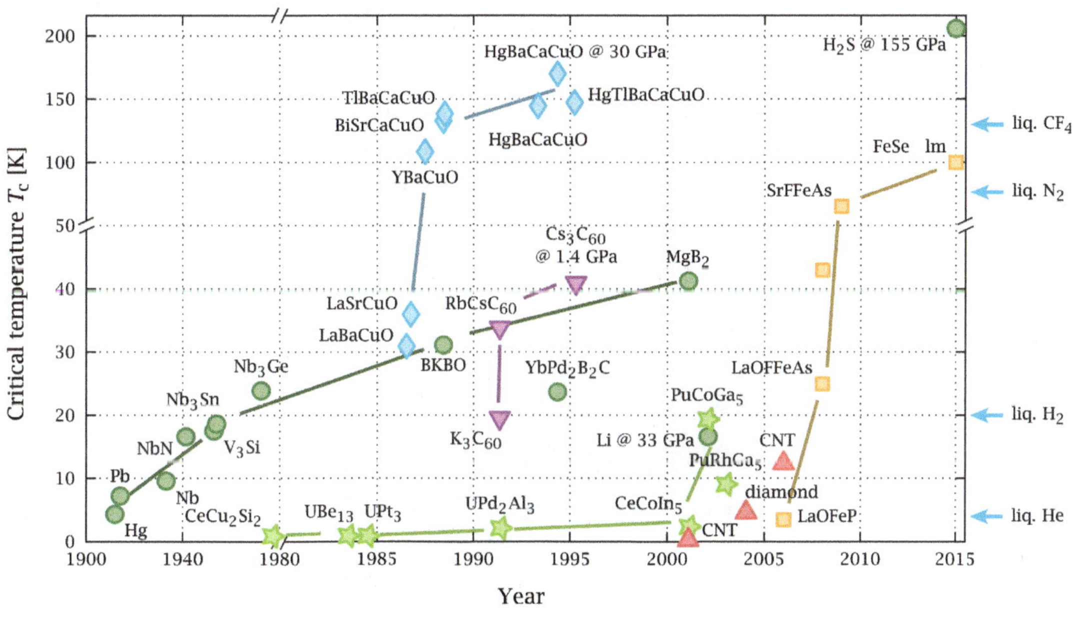

Timeline of superconducting materials

Until 1986, physicists had believed that BCS theory forbade superconductivity at temperatures above about 30 K. In that year, Bednorz and Müller discovered superconductivity in a lanthanum-based cuprate perovskite material, which had a transition temperature of 35 K (Nobel Prize in Physics, 1987). It was soon found that replacing the lanthanum with yttrium (i.e., making YBCO) raised the critical temperature to 92 K.

This temperature jump is particularly significant, since it allows liquid nitrogen as a refrigerant, replacing liquid helium. This can be important commercially because liquid nitrogen can be produced relatively cheaply, even on-site. Also, the higher temperatures help avoid some of the problems that arise at liquid helium temperatures, such as the formation of plugs of frozen air that can block cryogenic lines and cause unanticipated and potentially hazardous pressure buildup.

Many other cuprate superconductors have since been discovered, and the theory of superconductivity in these materials is one of the major outstanding challenges of theoretical condensed matter physics. There are currently two main hypotheses – the resonating-valence-bond theory, and spin fluctuation which has the most support in the research community. The second hypothesis proposed that electron pairing in high-temperature superconductors is mediated by short-range spin waves known as paramagnons.

Since about 1993, the highest-temperature superconductor has been a ceramic material consisting of mercury, barium, calcium, copper and oxygen ($HgBa_2Ca_2Cu_3O_{8+\delta}$) with T_c = 133–138 K. The latter experiment (138 K) still awaits experimental confirmation, however.

In February 2008, an iron-based family of high-temperature superconductors was discovered. Hideo Hosono, of the Tokyo Institute of Technology, and colleagues found lanthanum oxygen fluorine iron arsenide ($LaO_{1-x}F_xFeAs$), an oxypnictide that superconducts below 26 K. Replacing the lanthanum in $LaO_{1-x}F_xFeAs$ with samarium leads to superconductors that work at 55 K.

In May 2014, hydrogen sulfide (H_2S) was predicted to be a high-temperature superconductor with a transition temperature of 80 K at 160 gigapascals of pressure. In 2015, H_2S has been observed to exhibit superconductivity at below 203 K but at extremely high pressures — around 150 gigapascals.

Applications

Video of superconducting levitation of YBCO.

Superconducting magnets are some of the most powerful electromagnets known. They are used in MRI/NMR machines, mass spectrometers, the beam-steering magnets used in particle accelerators and plasma confining magnets in some tokamaks. They can also be used for magnetic separation, where weakly magnetic particles are extracted from a background of less or non-magnetic particles, as in the pigment industries.

In the 1950s and 1960s, superconductors were used to build experimental digital computers using cryotron switches. More recently, superconductors have been used to make digital circuits based on rapid single flux quantum technology and RF and microwave filters for mobile phone base stations.

Superconductors are used to build Josephson junctions which are the building blocks of SQUIDs (superconducting quantum interference devices), the most sensitive magnetometers known. SQUIDs are used in scanning SQUID microscopes and magnetoencephalography. Series of Josephson devices are used to realize the SI volt. Depending on the particular mode of operation, a superconductor-insulator-superconductor Josephson junction can be used as a photon detector or as a mixer. The large resistance change at the transition from the normal- to the superconducting state is used to build thermometers in cryogenic micro-calorimeter photon detectors. The same effect is used in ultrasensitive bolometers made from superconducting materials.

Other early markets are arising where the relative efficiency, size and weight advantages of devices based on high-temperature superconductivity outweigh the additional costs involved. For example, in wind turbines the lower weight and volume of superconducting generators could lead to savings in construction and tower costs, offsetting the higher costs for the generator and lowering the total LCOE.

Promising future applications include high-performance smart grid, electric power transmission, transformers, power storage devices, electric motors (e.g. for vehicle propulsion, as in vactrains or maglev trains), magnetic levitation devices, fault current limiters, enhancing spintronic devices with superconducting materials, and superconducting magnetic refrigeration. However, superconductivity is sensitive to moving magnetic fields so applications that use alternating current (e.g. transformers) will be more difficult to develop than those that rely upon direct current. Compared to traditional power lines superconducting transmission lines are more efficient and require only a fraction of the space, which would not only lead to a better environmental performance but could also improve public acceptance for expansion of the electric grid.

NOBEL PRIZES FOR SUPERCONDUCTIVITY

- Heike Kamerlingh Onnes (1913), "for his investigations on the properties of matter at low temperatures which led, inter alia, to the production of liquid helium"
- John Bardeen, Leon N. Cooper, and J. Robert Schrieffer (1972), "for their jointly developed theory of superconductivity, usually called the BCS-theory"
- Leo Esaki, Ivar Giaever, and Brian D. Josephson (1973), "for their experimental discoveries regarding tunneling phenomena in semiconductors and superconductors, respectively," and "for his theoretical predictions of the properties of a supercurrent through a tunnel barrier, in particular those phenomena which are generally known as the Josephson effects"
- Georg Bednorz and K. Alex Müller (1987), "for their important break-through in the discovery of superconductivity in ceramic materials"

- Alexei A. Abrikosov, Vitaly L. Ginzburg, and Anthony J. Leggett (2003), "for pioneering contributions to the theory of superconductors and superfluids"
- Michael Kosterlitz, Duncan Haldane, David J. Thouless (2016)

HIGH-TEMPERATURE SUPERCONDUCTIVITY

A small sample of the high-temperature superconductor BSCCO-2223.

High-temperature superconductors (abbreviated **high-T_c** or **HTS**) are materials that behave as superconductors at unusually high temperatures. The first high-T_c superconductor was discovered in 1986 by IBM researchers Georg Bednorz and K. Alex Müller, who were awarded the 1987 Nobel Prize in Physics "for their important break-through in the discovery of superconductivity in ceramic materials".

Whereas "ordinary" or metallic superconductors usually have transition temperatures (temperatures below which they are superconductive) below 30 K (−243.2 °C), and must be cooled using liquid helium in order to achieve superconductivity, HTS have been observed with transition temperatures as high as 138 K (−135 °C), and can be cooled to superconductivity using liquid nitrogen. Until 2008, only certain compounds of copper and oxygen (so-called "cuprates") were believed to have HTS properties, and the term high-temperature superconductor was used interchangeably with cuprate superconductor for compounds such as bismuth strontium calcium copper oxide (BSCCO) and yttrium barium copper oxide (YBCO). Several iron-based compounds (the iron pnictides) are now known to be superconducting at high temperatures.

In 2015, hydrogen sulfide (H_2S) under extremely high pressure (around 150 gigapascals) was found to undergo superconducting transition near 203 K (-70 °C), the highest temperature superconductor known to date.

For an explanation about T_c (the critical temperature for superconductivity), see Superconductivity § Superconducting phase transition and the second bullet item of BCS theory § Successes of the BCS theory.

The phenomenon of superconductivity was discovered by Kamerlingh Onnes in 1911, in metallic mercury below 4 K (−269.15 °C). For seventy-five years after that, researchers attempted to observe superconductivity at higher and higher temperatures. In the late 1970s, superconductivity was observed in certain metal oxides at temperatures as high as 13 K (−260.1 °C), which were much higher than those for elemental metals. In 1986, J. Georg Bednorz and K. Alex Müller, working at the IBM research lab near Zurich, Switzerland were exploring a new class of ceramics for superconductivity. Bednorz encountered a barium-doped compound of lanthanum and copper oxide whose resistance dropped down to zero at a temperature around 35 K (−238.2 °C). Their results were soon confirmed by many groups, notably Paul Chu at the University of Houston and Shoji Tanaka at the University of Tokyo.

Shortly after, P. W. Anderson, at Princeton University came up with the first theoretical description of these materials, using the resonating valence bond theory, but a full understanding of these materials is still developing today. These superconductors are now known to possess a *d*-wave pair symmetry. The first proposal that high-temperature cuprate superconductivity involves *d*-wave pairing was made in 1987 by Bickers, Scalapino and Scalettar, followed by three subsequent theories in 1988 by Inui, Doniach, Hirschfeld and Ruckenstein, using spin-fluctuation theory, and by Gros, Poilblanc, Rice and Zhang, and by Kotliar and Liu identifying *d*-wave pairing as a natural consequence of the RVB theory. The confirmation of the *d*-wave nature of the cuprate superconductors was made by a variety of experiments, including the direct observation of the *d*-wave nodes in the excitation spectrum through Angle Resolved Photoemission Spectroscopy, the observation of a half-integer flux in tunneling experiments, and indirectly from the temperature dependence of the penetration depth, specific heat and thermal conductivity.

The superconductor with the highest transition temperature that has been confirmed by multiple independent research groups (a prerequisite to be called a discovery, verified by peer review) is mercury barium calcium copper oxide ($HgBa_2Ca_2Cu_3O_8$) at around 133 K.

After more than twenty years of intensive research, the origin of high-temperature superconductivity is still not clear, but it seems that instead of *electron-phonon* attraction mechanisms, as in conventional superconductivity, one is dealing with genuine *electronic* mechanisms (e.g. by antiferromagnetic correlations), and instead of conventional, purely *s*-wave pairing, more exotic pairing symmetries are thought to be involved (*d*-wave in the case of the cuprates; primarily extended *s*-wave, but occasionally *d*-wave, in the case of the iron-based superconductors). One goal of all this research is room-temperature superconductivity. In 2014, evidence showing that fractional particles can happen in quasi two-dimensional magnetic materials, was found by EPFL scientists lending support for Anderson's theory of high-temperature superconductivity.

Crystal structures of high-temperature ceramic superconductors

The structure of high-T_c copper oxide or cuprate superconductors are often closely related to perovskite structure, and the structure of these compounds has been described as a distorted, oxygen deficient multi-layered perovskite structure. One of the properties of the crystal structure of oxide superconductors is an alternating multi-layer of CuO_2 planes with superconductivity taking place between these layers. The more layers of

CuO_2, the higher T_c. This structure causes a large anisotropy in normal conducting and superconducting properties, since electrical currents are carried by holes induced in the oxygen sites of the CuO_2 sheets. The electrical conduction is highly anisotropic, with a much higher conductivity parallel to the CuO_2 plane than in the perpendicular direction. Generally, critical temperatures depend on the chemical compositions, cations substitutions and oxygen content. They can be classified as superstripes; i.e., particular realizations of superlattices at atomic limit made of superconducting atomic layers, wires, dots separated by spacer layers, that gives multiband and multigap superconductivity.

YBaCuO Superconductors

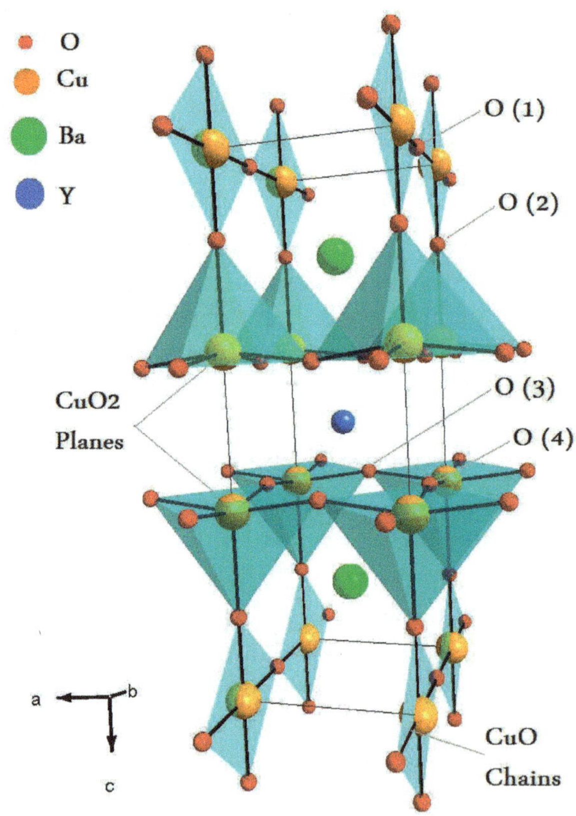

YBCO unit cell.

The first superconductor found with $T_c > 77$ K (liquid nitrogen boiling point) is yttrium barium copper oxide ($YBa_2Cu_3O_{7-x}$); the proportions of the three different metals in the $YBa_2Cu_3O_7$ superconductor are in the mole ratio of 1 to 2 to 3 for yttrium to barium to copper, respectively. Thus, this particular superconductor is often referred to as the 123 superconductor.

The unit cell of $YBa_2Cu_3O_7$ consists of three pseudocubic elementary perovskite unit cells. Each perovskite unit cell contains a Y or Ba atom at the center: Ba in the bottom unit cell, Y in the middle one, and Ba in the top unit cell. Thus, Y and Ba are stacked in the sequence [Ba–Y–Ba] along the c-axis. All corner sites of the unit cell are occupied by Cu, which has two different coordinations, Cu(1) and Cu(2), with respect to oxygen. There are four possible crystallographic sites for oxygen: O(1), O(2), O(3) and O(4). The coordination polyhedra of Y and Ba with respect to oxygen are different. The tripling of the perovskite unit cell leads to nine oxygen atoms, whereas $YBa_2Cu_3O_7$ has seven oxygen atoms and, therefore, is referred to as an oxygen-deficient perovskite structure. The structure has a stacking of different layers: (CuO)(BaO)(CuO_2)(Y)(CuO_2)(BaO)(CuO). One of the key feature of the unit cell of $YBa_2Cu_3O_{7-x}$ (YBCO) is the presence of two layers of CuO_2. The role of the Y plane is to serve as a spacer between two CuO_2 planes. In YBCO, the Cu–O chains are known to play an important role for superconductivity. T_c is maximal near 92 K when $x \approx 0.15$ and the structure is orthorhombic. Superconductivity disappears at $x \approx 0.6$, where the structural transformation of YBCO occurs from orthorhombic to tetragonal.

Bi-, Tl- and Hg-based high-T_c superconductors

The crystal structure of Bi-, Tl- and Hg-based high-T_c superconductors are very similar. Like YBCO, the perovskite-type feature and the presence of CuO_2 layers also exist in these superconductors. However, unlike YBCO, Cu–O chains are not present in these superconductors. The YBCO superconductor has an orthorhombic structure, whereas the other high-T_c superconductors have a tetragonal structure.

The **Bi–Sr–Ca–Cu–O** system has three superconducting phases forming a homologous series as $Bi_2Sr_2Ca_{n-1}Cu_nO_{4+2n+x}$ (n = 1, 2 and 3). These three phases are Bi-2201, Bi-2212 and Bi-2223, having transition temperatures of 20, 85 and 110 K, respectively, where the numbering system represent number of atoms for Bi, Sr, Ca and Cu respectively. The two phases have a tetragonal structure which consists of two sheared crystallographic unit cells. The unit cell of these phases has double Bi–O planes which are stacked in a way that the Bi atom of one plane sits below the oxygen atom of the next consecutive plane. The Ca atom forms a layer within the interior of the CuO_2 layers in both Bi-2212 and Bi-2223; there is no Ca layer in the Bi-2201 phase. The three phases differ with each other in the number of CuO_2 planes; Bi-2201, Bi-2212 and Bi-2223 phases have one, two and three CuO_2 planes, respectively. The c axis lattice constants of these phases increases with the number of CuO_2planes (see table below). The coordination of the Cu atom is different in the three phases. The Cu atom forms an octahedral coordination with respect to oxygen atoms in the 2201 phase, whereas in 2212, the Cu atom is surrounded by five oxygen atoms in a pyramidal arrangement. In the 2223 structure, Cu has two coordinations with respect to oxygen: one Cu atom is bonded with four oxygen atoms in square planar configuration and another Cu atom is coordinated with five oxygen atoms in a pyramidal arrangement.

Tl–Ba–Ca–Cu–O superconductor: The first series of the Tl-based superconductor containing one Tl–O layer has the general formula $TlBa_2Ca_{n-1}Cu_nO_{2n+3}$, whereas the second series containing two Tl–O layers has a formula of $Tl_2Ba_2Ca_{n-1}Cu_nO_{2n+4}$ with n = 1, 2 and 3. In the structure of $Tl_2Ba_2CuO_6$ (Tl-2201), there is one CuO_2 layer with the stacking sequence (Tl–O) (Tl–O) (Ba–O) (Cu–O) (Ba–O) (Tl–O) (Tl–O). In $Tl_2Ba_2CaCu_2O_8$(Tl-2212), there are two Cu–O layers with a Ca layer in between. Similar to the $Tl_2Ba_2CuO_6$ structure, Tl–O layers are present outside the Ba–O layers. In $Tl_2Ba_2Ca_2Cu_3O_{10}$ (Tl-2223), there are three CuO2 layers enclosing Ca layers between each of these. In Tl-based superconductors, T_c is found to increase with the increase in CuO_2 layers. However, the value of T_c decreases after four CuO_2 layers in $TlBa_2Ca_{n-1}Cu_nO_{2n+3}$, and in the $Tl_2Ba_2Ca_{n-1}Cu_nO_{2n+4}$ compound, it decreases after three CuO_2 layers.

Hg–Ba–Ca–Cu–O superconductor: The crystal structure of $HgBa_2CuO_4$ (Hg-1201), $HgBa_2CaCu_2O_6$ (Hg-1212) and $HgBa_2Ca_2Cu_3O_8$ (Hg-1223) is similar to that of Tl-1201, Tl-1212 and Tl-1223, with Hg in place of Tl. It is noteworthy that the T_c of the Hg compound (Hg-1201) containing one CuO_2 layer is much larger as compared to the one-CuO_2-layer compound of thallium (Tl-1201). In the Hg-based superconductor, T_c is also found to increase as the CuO_2 layer increases. For Hg-1201, Hg-1212 and Hg-1223, the values of T_c are 94, 128 and the record value at ambient pressure 134 K, respectively, as shown in table below. The observation that the T_c of Hg-1223 increases to 153 K under high pressure indicates that the T_c of this compound is very sensitive to the structure of the compound.

Critical temperature (T_c), crystal structure and lattice constants of some high-T_c superconductors

Formula	Notation	T_c (K)	No. of Cu-O planes in unit cell	Crystal structure
$YBa_2Cu_3O_7$	123	92	2	Orthorhombic
$Bi_2Sr_2CuO_6$	Bi-2201	20	1	Tetragonal
$Bi_2Sr_2CaCu_2O_8$	Bi-2212	85	2	Tetragonal
$Bi_2Sr_2Ca_2Cu_3O_{10}$	Bi-2223	110	3	Tetragonal
$Tl_2Ba_2CuO_6$	Tl-2201	80	1	Tetragonal
$Tl_2Ba_2CaCu_2O_8$	Tl-2212	108	2	Tetragonal
$Tl_2Ba_2Ca_2Cu_3O_{10}$	Tl-2223	125	3	Tetragonal
$TlBa_2Ca_3Cu_4O_{11}$	Tl-1234	122	4	Tetragonal
$HgBa_2CuO_4$	Hg-1201	94	1	Tetragonal
$HgBa_2CaCu_2O_6$	Hg-1212	128	2	Tetragonal
$HgBa_2Ca_2Cu_3O_8$	Hg-1223	134	3	Tetragonal

Preparation of High-T_c Superconductors

The simplest method for preparing high-T_c superconductors is a solid-state thermochemical reaction involving mixing, calcination and sintering. The appropriate amounts of precursor powders, usually oxides and carbonates, are mixed thoroughly using a Ball mill. Solution chemistry processes such as coprecipitation, freeze-drying and sol-gel methods are alternative ways for preparing a homogeneous mixture. These powders are calcined in the temperature range from 800 °C to 950 °C for several hours. The powders are cooled, reground and calcined again. This process is repeated several times to get homogeneous material. The powders are subsequently compacted to pellets and sintered. The sintering environment such as temperature, annealing time, atmosphere and cooling rate play a very important role in getting good high-T_c superconducting materials. The $YBa_2Cu_3O_{7-x}$ compound is prepared by calcination and sintering of a homogeneous mixture of Y_2O_3, $BaCO_3$ and CuO in the appropriate atomic ratio. Calcination is done at 900–950 °C, whereas sintering is done at 950 °C in an oxygen atmosphere. The oxygen stoichiometry in this material is very crucial for obtaining a superconducting $YBa_2Cu_3O_{7-x}$ compound. At the time of sintering, the semiconducting tetragonal $YBa_2Cu_3O_6$ compound is formed, which, on slow cooling in oxygen atmosphere, turns into superconducting $YBa_2Cu_3O_{7-x}$. The uptake and loss of oxygen are reversible in $YBa_2Cu_3O_{7-x}$. A fully oxygenated orthorhombic $YBa_2Cu_3O_{7-x}$ sample can be transformed into tetragonal $YBa_2Cu_3O_6$ by heating in a vacuum at temperature above 700 °C.

The preparation of Bi-, Tl- and Hg-based high-T_c superconductors is difficult compared to YBCO. Problems in these superconductors arise because of the existence of three or more phases having a similar layered structure. Thus, syntactic intergrowth and defects such as stacking faults occur during synthesis and it becomes difficult to isolate a single superconducting phase. For Bi–Sr–Ca–Cu–O, it is relatively simple to prepare the Bi-2212 ($T_c \approx 85$ K) phase, whereas it is very difficult to prepare a single phase of Bi-2223 ($T_c \approx 110$ K). The Bi-2212 phase appears only after few hours of sintering at 860–870 °C, but the larger fraction of the Bi-2223 phase is formed after a long reaction time of more than a week at 870 °C. Although the substitution of Pb in the Bi–Sr–Ca–Cu–O compound has been found to promote the growth of the high-T_c phase, a long sintering time is still required.

Properties

"High-temperature" has two common definitions in the context of superconductivity:

- Above the temperature of 30 K that had historically been taken as the upper limit allowed by BCS theory(1957). This is also above the 1973 record of 23 K that had lasted until copper-oxide materials were discovered in 1986.
- Having a transition temperature that is a larger fraction of the Fermi temperature than for conventional superconductors such as elemental mercury or lead. This definition encompasses a wider variety of unconventional superconductors and is used in the context of theoretical models.

The label high-T_c may be reserved by some authors for materials with critical temperature greater than the boiling point of liquid nitrogen (77 K or −196 °C). However, a number of materials – including the original discovery and recently discovered pnictide superconductors – had critical temperatures below 77 K but are commonly referred to in publication as being in the high-T_c class.

Technological applications could benefit from both the higher critical temperature being above the boiling point of liquid nitrogen and also the higher critical magnetic field (and critical current density) at which superconductivity is destroyed. In magnet applications, the high critical magnetic field may prove more valuable than the high T_c itself. Some cuprates have an upper critical field of about 100 tesla. However, cuprate materials are brittle ceramics which are expensive to manufacture and not easily turned into wires or other useful shapes. Also, high-temperature superconductors do not form large, continuous superconducting domains, but only clusters of microdomains within which superconducting occurs. They are therefore unsuitable for applications requiring actual superconducted currents, such as magnets for magnetic resonance spectrometers.

After two decades of intense experimental and theoretical research, with over 100,000 published papers on the subject, several common features in the properties of high-temperature superconductors have been identified. As of 2011, no widely accepted theory explains their properties. Relative to conventional superconductors, such as elemental mercury or lead that are adequately explained by the BCS theory, cuprate superconductors (and other unconventional superconductors) remain distinctive. There also has been much debate as to high-temperature superconductivity coexisting with magnetic ordering in YBCO, iron-based superconductors, several ruthenocuprates and other exotic superconductors, and the search continues for other families of materials. HTS are Type-II superconductors, which allow magnetic fields to penetrate their interior in quantized units of flux, meaning that much higher magnetic fields are required to suppress superconductivity. The layered structure also gives a directional dependence to the magnetic field response.

Cuprates

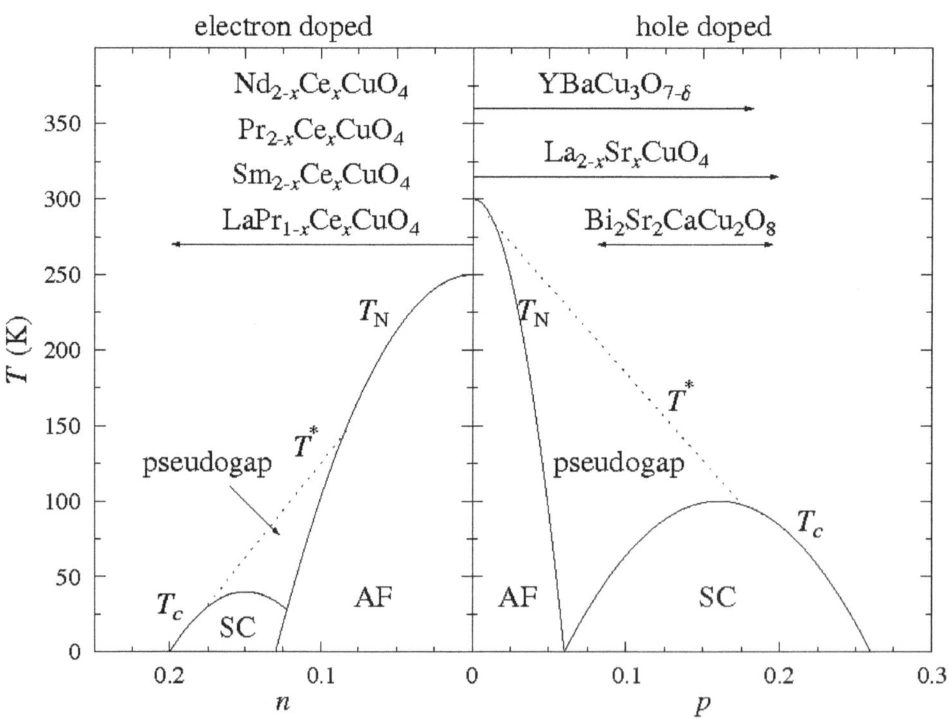

Simplified doping dependent phase diagram of cuprate superconductors for both electron (n) and hole (p) doping. The phases shown are the antiferromagnetic (AF) phase close to zero doping, the superconducting phase around optimal doping, and the pseudogap phase. Doping ranges possible for some common compounds are also shown. After.

Cuprate superconductors are generally considered to be quasi-two-dimensional materials with their superconducting properties determined by electrons moving within weakly coupled copper-oxide (CuO_2) layers. Neighbouring layers containing ions such as lanthanum, barium, strontium, or other atoms act to stabilize the structure and dope electrons or holes onto the copper-oxide layers. The undoped "parent" or "mother" compounds are Mott insulators with long-range antiferromagnetic order at low enough temperature. Single band models are generally considered to be sufficient to describe the electronic properties.

The cuprate superconductors adopt a perovskite structure. The copper-oxide planes are checkerboard lattices with squares of O^{2-} ions with a Cu^{2+} ion at the centre of each square. The unit cell is rotated by 45° from these squares. Chemical formulae of superconducting materials generally contain fractional numbers to describe the doping required for superconductivity. There are several families of cuprate superconductors and they can be categorized by the elements they contain and the number of adjacent copper-oxide layers in each superconducting block. For example, YBCO and BSCCO can alternatively be referred to as Y123 and Bi2201/Bi2212/Bi2223 depending on the number of layers in each superconducting block (n). The superconducting transition temperature has been found to peak at an optimal doping value ($p = 0.16$) and an optimal number of layers in each superconducting block, typically $n = 3$.

Possible mechanisms for superconductivity in the cuprates are still the subject of considerable debate and further research. Certain aspects common to all materials have been identified. Similarities between the antiferromagnetic low-temperature state of the undoped materials and the superconducting state that emerges upon doping, primarily the $d_{x^2-y^2}$ orbital state of the Cu^{2+} ions, suggest that electron-electron interactions are more significant than electron-phonon interactions in cuprates – making the superconductivity unconventional. Recent work on the Fermi surface has shown that nesting occurs at four points in the antiferromagnetic Brillouin zone where spin waves exist and that the superconducting energy gap is larger at these points. The weak isotope effects observed for most cuprates contrast with conventional superconductors that are well described by BCS theory.

Similarities and differences in the properties of hole-doped and electron doped cuprates:

- Presence of a pseudogap phase up to at least optimal doping.
- Different trends in the Uemura plot relating transition temperature to the superfluid density. The inverse square of the London penetration depth appears to be proportional to the critical temperature for a large number of underdoped cuprate superconductors, but the constant of proportionality is different for hole- and electron-doped cuprates. The linear trend implies that the physics of these materials is strongly two-dimensional.
- Universal hourglass-shaped feature in the spin excitations of cuprates measured using inelastic neutron diffraction.
- Nernst effect evident in both the superconducting and pseudogap phases.

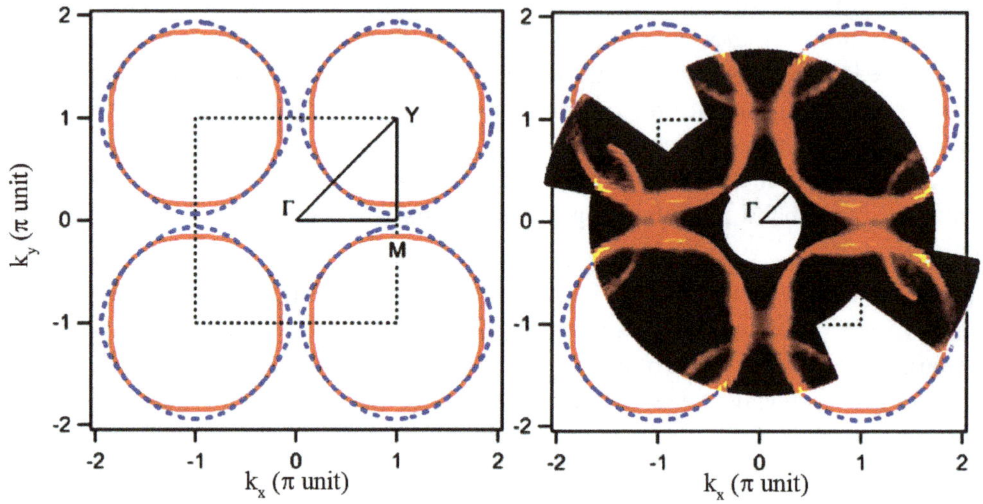

Fig. 1. The Fermi surface of bi-layer BSCCO, calculated (left) and measured by ARPES (right). The dashed rectangle represents the first Brillouin zone.

The electronic structure of superconducting cuprates is highly anisotropic (see the crystal structure of YBCO or BSCCO). Therefore, the Fermi surface of HTSC is very close to the Fermi surface of the doped CuO_2 plane (or multi-planes, in case of multi-layer cuprates) and can be presented on the 2D reciprocal space (or momentum space) of the CuO_2 lattice. The typical Fermi surface within the first CuO_2 Brillouin zone is sketched in Fig. 1 (left). It can be derived from the band structure calculations or measured by angle resolved photoemission spectroscopy (ARPES). Fig. 1 (right) shows the Fermi surface of BSCCO measured by ARPES. In a wide range of charge carrier concentration (doping level), in which the hole-doped HTSC are superconducting, the Fermi surface is hole-like (*i.e.* open, as shown in Fig. 1). This results in an inherent in-plane anisotropy of the electronic properties of HTSC.

Iron-based Superconductors

Iron-based superconductors contain layers of iron and a pnictogen—such as arsenic or phosphorus—or a chalcogen. This is currently the family with the second highest critical temperature, behind the cuprates. Interest in their superconducting properties began in 2006 with the discovery of superconductivity in LaFePO at 4 K and gained much greater attention in 2008 after the analogous material LaFeAs(O,F) was found to superconduct at up to 43 K under pressure. The highest critical temperatures in the iron-based superconductor family exist in thin films of FeSe, where a critical temperature in excess of 100 K has recently been reported.

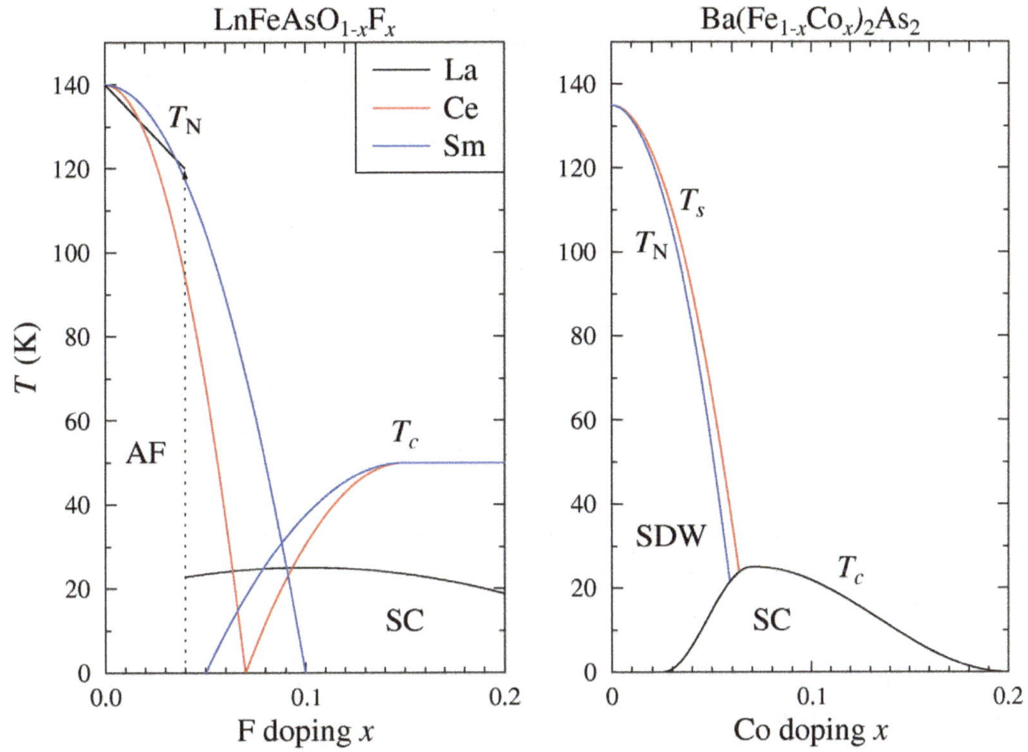

Simplified doping dependent phase diagrams of iron-based superconductors for both Ln-1111 and Ba-122 materials. The phases shown are the antiferromagnetic/spin density wave (AF/SDW) phase close to zero doping and the superconducting phase around optimal doping. The Ln-1111 phase diagrams for La and Smwere determined using muon spin spectroscopy, the phase diagram for Ce was determined using neutron diffraction. The Ba-122 phase diagram is based on.

Since the original discoveries several families of iron-based superconductors have emerged:

- LnFeAs(O,F) or $LnFeAsO_{1-x}$ (Ln = lanthanide) with T_c up to 56 K, referred to as 1111 materials. A fluoride variant of these materials was subsequently found with similar $T_c$values.
- $(Ba,K)Fe_2As_2$ and related materials with pairs of iron-arsenide layers, referred to as 122 compounds. T_c values range up to 38 K. These materials also superconduct when iron is replaced with cobalt
- LiFeAs and NaFeAs with T_c up to around 20 K. These materials superconduct close to stoichiometric composition and are referred to as 111 compounds.
- FeSe with small off-stoichiometry or tellurium doping.

Most undoped iron-based superconductors show a tetragonal-orthorhombic structural phase transition followed at lower temperature by magnetic ordering, similar to the cuprate superconductors. However, they are poor metals rather than Mott insulators and have five bands at the Fermi surface rather than one. The phase diagram emerging as the iron-arsenide layers are doped is remarkably similar, with the superconducting phase close to or overlapping the magnetic phase. Strong evidence that the T_c value varies with the As-Fe-As bond angles has already emerged and shows that the optimal T_c value is obtained with undistorted $FeAs_4$ tetrahedra. The symmetry of the pairing wavefunction is still widely debated, but an extended *s*-wave scenario is currently favoured.

Hydrogen Sulfide

At pressures above 90 GPa (Gigapascals), hydrogen sulfide becomes a metallic conductor of electricity. When cooled below a critical temperature this high-pressure phase exhibits superconductivity. The critical temperature increases with pressure, ranging from 23 K at 100 GPa to 150 K at 200 GPa. If hydrogen sulfide is pressurized at higher temperatures, then cooled, the critical temperature reaches 203 K (−70 °C), the highest accepted superconducting critical temperature as of 2015. It has been predicted that by substituting a small part of sulfur with phosphorus and using even higher pressures it may be possible to raise the critical temperature to above 0 °C (273 K) and achieve room-temperature superconductivity.

Other Materials Sometimes Referred to as High-Temperature Superconductors

Magnesium diboride is occasionally referred to as a high-temperature superconductor because its T_c value of 39 K is above that historically expected for BCS superconductors. However, it is more generally regarded as the highest-T_c conventional superconductor, the increased T_c resulting from two separate bands being present at the Fermi level.

Fulleride superconductors where alkali-metal atoms are intercalated into C_{60} molecules show superconductivity at temperatures of up to 38 K for Cs_3C_{60}.

Some organic superconductors and heavy fermion compounds are considered to be high-temperature superconductors because of their high T_c values relative to their Fermi energy, despite the T_c values being lower than for many conventional superconductors. This description may relate better to common aspects of the superconducting mechanism than the superconducting properties.

Metallic Hydrogen

Theoretical work by Neil Ashcroft in 1968 predicted that solid metallic hydrogen at extremely high pressure should become superconducting at approximately room-temperature because of its extremely high speed of sound and expected strong coupling between the conduction electrons and the lattice vibrations. This prediction is yet to be experimentally verified.

Magnetic Properties

All known high-T_c superconductors are Type-II superconductors. In contrast to Type-I superconductors, which expel all magnetic fields due to the Meissner effect, Type-II superconductors allow magnetic fields to penetrate their interior in quantized units of flux, creating "holes" or "tubes" of normal metallic regions in the superconducting bulk called vortices. Consequently, high-T_c superconductors can sustain much higher magnetic fields.

Ongoing Research

The question of how superconductivity arises in high-temperature superconductors is one of the major unsolved problems of theoretical condensed matter physics. The mechanism that causes the electrons in these crystals to form pairs is not known. Despite intensive research and many promising leads, an explanation has so far eluded scientists. One reason for this is that the materials in question are generally very complex, multi-layered crystals (for example, BSCCO), making theoretical modelling difficult.

Improving the quality and variety of samples also gives rise to considerable research, both with the aim of improved characterisation of the physical properties of existing compounds, and synthesizing new materials, often with the hope of increasing T_c. Technological research focuses on making HTS materials in sufficient quantities to make their use economically viable and optimizing their properties in relation to applications.

Possible Mechanism

There have been two representative theories for high-temperature or unconventional superconductivity. Firstly, **weak coupling theory** suggests superconductivity emerges from antiferromagnetic spin fluctuations in a doped system. According to this theory, the pairing wave function of the cuprate HTS should have a $d_{x^2-y^2}$ symmetry. Thus, determining whether the pairing wave function has *d*-wave symmetry is essential to test the spin fluctuation mechanism. That is, if the HTS order parameter (pairing wave function) does not have *d*-wave symmetry, then a pairing mechanism related to spin fluctuations can be ruled out. (Similar arguments can be made for iron-based superconductors but the different material properties allow a different pairing symmetry.) Secondly, there was the **interlayer coupling model**, according to which a layered structure consisting of BCS-type (*s*-wave symmetry) superconductors can enhance the superconductivity by itself. By introducing an additional tunnelling interaction between each layer, this model successfully explained the anisotropic symmetry of the order parameter as well as the emergence of the HTS. Thus, in order to solve this unsettled problem, there have been numerous experiments such as photoemission spectroscopy, NMR, specific heat measurements, etc. Up to date the results were ambiguous, some reports supported the *d* symmetry for the HTS whereas others supported the *s* symmetry. This muddy situation possibly originated from the indirect nature of the experimental evidence, as well as experimental issues such as sample quality, impurity scattering, twinning, etc.

Junction Experiment Supporting the d Symmetry

An experiment based on flux quantization of a three-grain ring of $YBa_2Cu_3O_7$ (YBCO) was proposed to test the symmetry of the order parameter in the HTS. The symmetry of the order parameter could best be probed at the junction interface as the Cooper pairs tunnel across a Josephson junction or weak link. It was expected that a half-integer flux, that is, a spontaneous magnetization could only occur for a junction of *d* symmetry superconductors. But, even if the junction experiment is the strongest method to determine the symmetry of the HTS order parameter, the results have been ambiguous. J. R. Kirtley and C. C. Tsuei thought that the ambiguous results came from the defects inside the HTS, so that they designed an experiment where both clean limit (no defects) and dirty limit (maximal defects) were considered simultaneously. In the experiment, the spontaneous magnetization was clearly observed in YBCO, which supported the *d* symmetry of the order parameter in YBCO. But, since YBCO is orthorhombic, it might inherently have an admixture of *s* symmetry. So, by tuning their technique further, they found that there was an admixture of *s* symmetry in YBCO within about 3%. Also, they found that there was a pure $d_{x^2-y^2}$ order parameter symmetry in the tetragonal $Tl_2Ba_2CuO_6$.

Qualitative Explanation of the Spin-Fluctuation Mechanism

Despite all these years, the mechanism of high-T_c superconductivity is still highly controversial, mostly due to the lack of exact theoretical computations on such strongly interacting electron systems. However, most rigorous theoretical calculations, including phenomenological and diagrammatic approaches, converge on magnetic fluctuations as the pairing mechanism for these systems. The qualitative explanation is as follows:

In a superconductor, the flow of electrons cannot be resolved into individual electrons, but instead consists of many pairs of bound electrons, called Cooper pairs. In conventional superconductors, these pairs are formed when an electron moving through the material distorts the surrounding crystal lattice, which in turn attracts another electron and forms a bound pair. This is sometimes called the "water bed" effect. Each Cooper pair requires a certain minimum energy to be displaced, and if the thermal fluctuations in the crystal lattice are smaller than this energy the pair can flow without dissipating energy. This ability of the electrons to flow without resistance leads to superconductivity.

In a high-T_c superconductor, the mechanism is extremely similar to a conventional superconductor, except, in this case, phonons virtually play no role and their role is replaced by spin-density waves. Just as all known conventional superconductors are strong phonon systems, all known high-T_c superconductors are strong spin-density wave systems, within close vicinity of a magnetic transition to, for example, an antiferromagnet. When an electron moves in a high-T_c superconductor, its spin creates a spin-density wave around it. This spin-density wave in turn causes a nearby electron to fall into the spin depression created by the first electron (water-bed effect again). Hence, again, a Cooper pair is formed. When the system temperature is lowered, more spin density waves and Cooper pairs are created, eventually leading to superconductivity. Note that in high-T_c systems, as these systems are magnetic systems due to the Coulomb interaction, there is a strong Coulomb repulsion between electrons. This Coulomb repulsion prevents pairing of the Cooper pairs on the same lattice site. The pairing of the electrons occur at near-neighbor lattice sites as a result. This is the so-called *d*-wave pairing, where the pairing state has a node (zero) at the origin.

Examples

Examples of high-T_c cuprate superconductors include $La_{1.85}Ba_{0.15}CuO_4$, and YBCO (Yttrium-Barium-Copper-Oxide), which is famous as the first material to achieve superconductivity above the boiling point of liquid nitrogen.

Transition temperatures of well-known superconductors (Boiling point of liquid nitrogen for comparison)			
Transition temperature (in kelvin)	**Transition temperature (in Celsius)**	**Material**	**Class**
203	−70	H_2S (at 150 GPa pressure)	Hydrogen-based superconductor
195	−78	Sublimation point of dry ice	
184	−89.2	Lowest temperature recorded on Earth	
133	−140	$HgBa_2Ca_2Cu_3O_x$(HBCCO)	Copper-oxide superconductors
110	−163	$Bi_2Sr_2Ca_2Cu_3O_{10}$(BSCCO)	
93	−180	$YBa_2Cu_3O_7$ (YBCO)	
90	−183	Boiling point of liquid oxygen	
77	−196	Boiling point of liquid nitrogen	
55	−218	SmFeAs(O,F)	Iron-based superconductors
41	−232	CeFeAs(O,F)	
26	−247	LaFeAs(O,F)	
20	−253	Boiling point of liquid hydrogen	
18	−255	Nb_3Sn	Metallic low-temperature superconductors
10	−263	NbTi	
9.2	−263.8	Nb	
4.2	−268.8	Boiling point of liquid helium	
4.2	−268.8	Hg (mercury)	Metallic low-temperature superconductors

List of Superconductors

The table showing major parameters of major superconductors of simple structure. X:Y means material X doped with element Y, T_C is the highest reported transition temperature in kelvins and H_C is a critical magnetic field in teslas. "BCS" means whether or not the superconductivity is explained within the BCS theory.

Formula	T_C (K)	H_C (T)	Type	BCS
Elements				
Al	1.20	0.01	I	yes
Cd	0.52	0.0028	I	yes
Diamond:B	11.4	4	II	yes

Ga	1.083	0.0058	I	yes
Hf	0.165		I	yes
α-Hg	4.15	0.04	I	yes
β-Hg	3.95	0.04	I	yes
In	3.4	0.03	I	yes
Ir	0.14	0.0016	I	yes
α-La	4.9		I	yes
β-La	6.3		I	yes
Mo	0.92	0.0096	I	yes
Nb	9.26	0.82	II	yes
Os	0.65	0.007	I	yes
Pa	1.4		I	yes
Pb	7.19	0.08	I	yes
Re	2.4	0.03	I	yes
Ru	0.49	0.005	I	yes
Si:B	0.4	0.4	II	yes
Sn	3.72	0.03	I	yes
Ta	4.48	0.09	I	yes
Tc	7.46–11.2	0.04	II	yes
α-Th	1.37	0.013	I	yes
Ti	0.39	0.01	I	yes
Tl	2.39	0.02	I	yes
α-U	0.68		I	yes
β-U	1.8		I	yes
V	5.03	1	II	yes
α-W	0.015	0.00012	I	yes
β-W	1–4			
Zn	0.855	0.005	I	yes
Zr	0.55	0.014	I	yes
Compounds				
Ba_8Si_{46}	8.07	0.008	II	yes
C_6Ca	11.5	0.95	II	
$C_6Li_3Ca_2$	11.15		II	
C_8K	0.14		II	
C_8KHg	1.4		II	
C_6K	1.5		II	
C_3K	3.0		II	
C_3Li	<0.35		II	
C_2Li	1.9		II	
C_3Na	2.3–3.8		II	
C_2Na	5.0		II	
C_8Rb	0.025		II	

C_6Sr	1.65		II	
C_6Yb	6.5		II	
$C_{60}Cs_2Rb$	33		II	yes
$C_{60}K_3$	19.8	0.013	II	yes
$C_{60}Rb_X$	28		II	yes
FeB_4	2.9		I	
InN	3		II	yes
In_2O_3	3.3	~3	II	yes
LaB_6	0.45			yes
MgB_2	39	74	II	yes
Nb_3Al	18		II	yes
Nb_3Ge	23.2	37	II	yes
NbO	1.38		II	yes
NbN	16		II	yes
Nb_3Sn	18.3	30	II	yes
NbTi	10	15	II	yes
SiC:B	1.4	0.008	I	yes
SiC:Al	1.5	0.04	II	yes
TiN	5.6			yes
YB_6	8.4		II	yes
ZrN	10			yes
ZrB_{12}	6.0		I	yes

REFERENCES

1. John Bardeen; Leon Cooper; J. R. Schriffer (December 1, 1957). "Theory of Superconductivity". Physical Review. 8 (5): 1178. Bibcode:1957PhRv..108.1175B. doi:10.1103/physrev.108.1175. ISBN 9780677000800. Retrieved June 6, 2014. reprinted in Nikolaĭ Nikolaevich Bogoliubov (1963) The Theory of Superconductivity, Vol. 4, CRC Press, ISBN 0677000804, p. 73
2. John Daintith (2009). The Facts on File Dictionary of Physics (4th ed.). Infobase Publishing. p. 238. ISBN 1438109490.
3. John C. Gallop (1990). SQUIDS, the Josephson Effects and Superconducting Electronics. CRC Press. pp. 3, 20. ISBN 0-7503-0051-5.
4. Durrant, Alan (2000). Quantum Physics of Matter. CRC Press. pp. 102–103. ISBN 0750307218.
5. J. G. Bednorz & K. A. Müller (1986). "Possible high Tcsuperconductivity in the Ba−La−Cu−O system". Z. Phys. B. 64 (1): 189–193. Bibcode:1986ZPhyB..64..189B. doi:10.1007/BF01303701.
6. Jun Nagamatsu; Norimasa Nakagawa; Takahiro Muranaka; Yuji Zenitani; et al. (2001). "Superconductivity at 39 K in magnesium diboride". Nature. 410 (6824):

63–4. Bibcode:2001Natur.410...63N. doi:10.1038/35065039. PMID 11242039.

7. Paul Preuss (14 August 2002). "A most unusual superconductor and how it works: first-principles calculation explains the strange behavior of magnesium diboride". Research News. Lawrence Berkeley National Laboratory. Retrieved 2009-10-28.
8. Hamish Johnston (17 February 2009). "Type-1.5 superconductor shows its stripes". Physics World. Institute of Physics. Retrieved 2009-10-28.
9. R. L. Dolecek (1954). "Adiabatic Magnetization of a Superconducting Sphere". Physical Review. 96 (1): 25–28. Bibcode:1954PhRv...96...25D. doi:10.1103/PhysRev.96.25.
10. H. Kleinert (1982). "Disorder Version of the Abelian Higgs Model and the Order of the Superconductive Phase Transition" (PDF). Lettere al Nuovo Cimento. 35 (13): 405–412. doi:10.1007/BF02754760.
11. J. Hove; S. Mo; A. Sudbo (2002). "Vortex interactions and thermally induced crossover from type-I to type-II superconductivity" (PDF). Physical Review B. 66 (6): 064524. arXiv:cond-mat/0202215 . Bibcode:2002PhRvB..66f4524H. doi:10.1103/PhysRevB.66.064524.
12. Lev D. Landau; Evgeny M. Lifschitz (1984). Electrodynamics of Continuous Media. Course of Theoretical Physics. 8. Oxford: Butterworth-Heinemann. ISBN 0-7506-2634-8.
13. David J. E. Callaway (1990). "On the remarkable structure of the superconducting intermediate state". Nuclear Physics B. 344 (3): 627–645. Bibcode:1990NuPhB.344..627C. doi:10.1016/0550-3213(90)90672-Z.
14. H. K. Onnes (1911). "The resistance of pure mercury at helium temperatures". Commun. Phys. Lab. Univ. Leiden. 12: 120.
15. Dirk vanDelft & Peter Kes (September 2010). "The Discovery of Superconductivity" (PDF). Physics Today. American Institute of Physics. 63 (9): 38–43. Bibcode:2010PhT....63i..38V. doi:10.1063/1.3490499.
16. W. Meissner & R. Ochsenfeld (1933). "Ein neuer Effekt bei Eintritt der Supraleitfähigkeit". Naturwissenschaften. 21 (44): 787–788. Bibcode:1933NW.....21..787M. doi:10.1007/BF01504252.
17. F. London & H. London (1935). "The Electromagnetic Equations of the Supraconductor". Proceedings of the Royal Society of London A. 149 (866): 71–88. Bibcode:1935RSPSA.149...71L. doi:10.1098/rspa.1935.0048. JSTOR 96265.
18. Meissner, W.; R. Ochsenfeld (1933). "Ein neuer Effekt bei Eintritt der Supraleitfähigkeit". Naturwissenschaften. 21 (44): 787–788. Bibcode:1933NW.....21..787M. doi:10.1007/BF01504252.
19. "The London equations". The Open University. Retrieved 2011-10-16.
20. J. Bardeen; L. N. Cooper & J. R. Schrieffer (1957). "Microscopic Theory of Superconductivity". Physical Review. 106 (1): 162–164. Bibcode:1957PhRv..106..162B. doi:10.1103/PhysRev.106.162.

21. J. Bardeen; L. N. Cooper & J. R. Schrieffer (1957). "Theory of Superconductivity". Physical Review. 108 (5): 1175–1205. Bibcode:1957PhRv..108.1175B. doi:10.1103/PhysRev.108.1175.

22. V. L. Ginzburg & L.D. Landau (1950). "On the theory of superconductivity". Zhurnal Eksperimental'noi i Teoreticheskoi Fiziki. 20: 1064.

23. E. Maxwell (1950). "Isotope Effect in the Superconductivity of Mercury". Physical Review. 78 (4): 477. Bibcode:1950PhRv...78..477M. doi:10.1103/PhysRev.78.477.

24. C. A. Reynolds; B. Serin; W. H. Wright & L. B. Nesbitt (1950). "Superconductivity of Isotopes of Mercury". Physical Review. 78 (4): 487. Bibcode:1950PhRv...78..487R. doi:10.1103/PhysRev.78.487.

25. N. N. Bogoliubov (1958). "A new method in the theory of superconductivity". Zhurnal Eksperimental'noi i Teoreticheskoi Fiziki. 34: 58.

26. L. P. Gor'kov (1959). "Microscopic derivation of the Ginzburg—Landau equations in the theory of superconductivity". Zhurnal Eksperimental'noi i Teoreticheskoi Fiziki. 36: 1364.

27. M. Combescot; W.V. Pogosov and O. Betbeder-Matibet (2013). "BCS ansatz for superconductivity in the light of the Bogoliubov approach and the Richardson–Gaudin exact wave function". Physica C: Superconductivity. 485: 47–57. arXiv:1111.4781 . Bibcode:2013PhyC..485...47C. doi:10.1016/j.physc.2012.10.011. Retrieved 11 August 2014.

28. Buck, Dudley A. "The Cryotron - A Superconductive Computer Component" (PDF). Lincoln Laboratory, Massachusetts Institute of Technology. Retrieved 10 August 2014.

29. G.B.Yntema (1955). "Superconducting Winding for Electromagnet". Physical Review. 98 (4): 1197. Bibcode:1955PhRv...98.1144.. doi:10.1103/PhysRev.98.1144.

30. J.E. Kunzler, E. Buehler, F.L.S. Hsu, and J.H. Wernick (1961). "Superconductivity in Nb3Sn at High Current Density in a Magnetic Field of 88 kgauss". Physical Review Letters. 6 (3): 89–91. Bibcode:1961PhRvL...6...89K. doi:10.1103/PhysRevLett.6.89.

31. T.G. Berlincourt and R.R. Hake (1962). "Pulsed-Magnetic-Field Studies of Superconducting Transition Metal Alloys at High and Low Current Densities". Bulletin of the American Physical Society. II–7: 408.

32. T.G. Berlincourt (1987). "Emergence of Nb-Ti as Supermagnet Material". Cryogenics. 27 (6): 283–289. Bibcode:1987Cryo...27..283B. doi:10.1016/0011-2275(87)90057-9.

33. B. D. Josephson (1962). "Possible new effects in superconductive tunnelling". Physics Letters. 1 (7): 251–253. Bibcode:1962PhL.....1..251J. doi:10.1016/0031-9163(62)91369-0.

34. "Newly discovered fundamental state of matter, a superinsulator, has been created.". Science Daily. April 9, 2008. Retrieved 2008-10-23.

35. M. K. Wu; et al. (1987). "Superconductivity at 93 K in a New Mixed-Phase

Y-Ba-Cu-O Compound System at Ambient Pressure". Physical Review Letters. 58 (9): 908–910. Bibcode:1987PhRvL..58..908W. doi:10.1103/PhysRevLett.58.908. PMID 10035069.

36. "Introduction to Liquid Helium". "Cryogenics and Fluid Branch". Goddard Space Flight Center, NASA.
37. "Section 4.1 "Air plug in the fill line"". "Superconducting Rock Magnetometer Cryogenic System Manual". 2G Enterprises. Archived from the original on May 6, 2009. Retrieved 9 October 2012.
38. Alexei A. Abrikosov (8 December 2003). "type II Superconductors and the Vortex Lattice". Nobel Lecture.
39. Adam Mann (Jul 20, 2011). "High-temperature superconductivity at 25: Still in suspense". Nature. 475 (7356): 280–2. Bibcode:2011Natur.475..280M. doi:10.1038/475280a. PMID 21776057.
40. Pines, D. (2002), "The Spin Fluctuation Model for High Temperature Superconductivity: Progress and Prospects", The Gap Symmetry and Fluctuations in High-Tc Superconductors, NATO Science Series: B:, 371, New York: Kluwer Academic, pp. 111–142, doi:10.1007/0-306-47081-0_7, ISBN 0-306-45934-5
41. P. Monthoux; A. V. Balatsky & D. Pines (1991). "Toward a theory of high-temperature superconductivity in the antiferromagnetically correlated cuprate oxides". Phys. Rev. Lett. 67 (24): 3448–3451. Bibcode:1991PhRvL..67.3448M. doi:10.1103/PhysRevLett.67.3448. PMID 10044736.
42. A. Schilling; et al. (1993). "Superconductivity above 130 K in the Hg–Ba–Ca–Cu–O system". Nature. 363 (6424): 56–58. Bibcode:1993Natur.363...56S. doi:10.1038/363056a0.
43. P. Dai; B. C. Chakoumakos; G. F. Sun; K. W. Wong; et al. (1995). "Synthesis and neutron powder diffraction study of the superconductor HgBa2Ca2Cu3O8+δ by Tl substitution". Physica C. 243 (3–4): 201–206. Bibcode:1995PhyC..243..201D. doi:10.1016/0921-4534(94)02461-8.
44. Hiroki Takahashi; Kazumi Igawa; Kazunobu Arii; Yoichi Kamihara; et al. (2008). "Superconductivity at 43 K in an iron-based layered compound LaO1−xFxFeAs". Nature. 453 (7193): 376–378. Bibcode:2008Natur.453..376T. doi:10.1038/nature06972. PMID 18432191.
45. Adrian Cho. "Second Family of High-Temperature Superconductors Discovered". ScienceNOW Daily News.
46. Zhi-An Ren; et al. (2008). "Superconductivity and phase diagram in iron-based arsenic-oxides ReFeAsO1-d (Re = rare-earth metal) without fluorine doping". EPL. 83: 17002. arXiv:0804.2582 . Bibcode:2008EL.....8317002R. doi:10.1209/0295-5075/83/17002.
47. Li, Yinwei; Hao, Jian; Liu, Hanyu; Li, Yanling; Ma, Yanming (2014-05-07). "The metallization and superconductivity of dense hydrogen sulfide". The Journal of Chemical Physics. 140 (17): 174712. arXiv:1402.2721 . Bibcode:2014JChPh.140q4712L. doi:10.1063/1.4874158. ISSN 0021-9606. PMID 24811660.

48. Drozdov, A. P.; Eremets, M. I.; Troyan, I. A.; Ksenofontov, V.; Shylin, S. I. (2015). "Conventional superconductivity at 203 kelvin at high pressures in the sulfur hydride system". Nature. 525 (7567): 73–6. arXiv:1506.08190 . Bibcode:2015Natur.525...73D. doi:10.1038/nature14964. ISSN 0028-0836. PMID 26280333.

49. Islam; et al. (2014). "A review of offshore wind turbine nacelle: Technical challenges, and research and developmental trends.". Renewable and Sustainable Energy Reviews. 33: 161–176. doi:10.1016/j.rser.2014.01.085.

50. Linder, Jacob; Robinson, Jason W. A. (2 April 2015). "Superconducting spintronics". Nature Physics. 11 (4): 307–315. arXiv:1510.00713 . Bibcode:2015NatPh..11..307L. doi:10.1038/nphys3242.

51. Thomas; et al. (2016). "Superconducting transmission lines – Sustainable electric energy transfer with higher public acceptance?". Renewable and Sustainable Energy Reviews. 55: 59–72. doi:10.1016/j.rser.2015.10.041.

52. "All Nobel Prizes in Physics". Nobelprize.org. Nobel Media AB 2014.

53. Timmer, John (May 2011). "25 years on, the search for higher-temp superconductors continues". Ars Technica. Retrieved 2 March 2012.

54. Ford, P. J. (2005). The Rise of the Superconductors. CRC Press.

55. Bednorz, J. G.; Müller, K. A. (1986). "Possible high TCsuperconductivity in the Ba-La-Cu-O system". Zeitschrift für Physik B. 64 (2): 189–193. Bibcode:1986ZPhyB..64..189B. doi:10.1007/BF01303701.

56. The Nobel Prize in Physics 1987: J. Georg Bednorz, K. Alex Müller. Nobelprize. org. Retrieved 2012-04-19.

57. Leggett (2006). "What DO we know about high Tc?". Nature Physics. 2 (3): 134–136. Bibcode:2006NatPh...2..134L. doi:10.1038/nphys254.

58. Choi, Charles Q. Iron Exposed as High-Temperature Superconductor: Scientific American. April 23, 2008. Retrieved 2012-04-19.

59. Ren, Zhi-An; Che, Guang-Can; Dong, Xiao-Li; Yang, Jie; Lu, Wei; Yi, Wei; Shen, Xiao Li; Li, Zheng Cai; Sun, Li-Ling; Zhou, Fang; Zhao, Zhong-Xian (2008). "Superconductivity and phase diagram in iron-based arsenic-oxides ReFeAsO1−δ (Re = rare-earth metal) without fluorine doping". EPL. 83: 17002. arXiv:0804.2582 . Bibcode:2008EL.....8317002R. doi:10.1209/0295-5075/83/17002.

60. Drozdov, A. P.; Eremets, M. I.; Troyan, I. A.; Ksenofontov, V.; Shylin, S. I. (2015). "Conventional superconductivity at 203 kelvin at high pressures in the sulfur hydride system". Nature. 525: 73–6. arXiv:1506.08190 . Bibcode:2015Natur.525...73D. doi:10.1038/nature14964. PMID 26280333.

61. Cartlidge, Edwin (18 August 2015). "Superconductivity record sparks wave of follow-up physics". Nature News. Retrieved 18 August 2015.

62. Christoph Heil; Lilia Boeri. "Influence of bonding on superconductivity in high-pressure hydrides". Phys Rev B. 92. arXiv:1507.02522 . Bibcode:2015PhRvB..92f0508H. doi:10.1103/PhysRevB.92.060508.

63. Nisbett, Alec (Producer) (1988). Superconductor: The race for the prize (Television Episode).

64. Stuart A Wolf & Vladimir Z Kresin,Eds, Novel Superconductivity, Springer (October, 1987)

65. Tanaka, Shoji (2001). "High temperature superconductivity: History and Outlook" (PDF). JSAP International. Retrieved 2 March 2012.

66. Anderson, Philip (1987). "The Resonating valence bond state in la-2CuO-4 and superconductivity". Science. 235 (4793): 1196–1198. Bibcode:1987Sci...235.1196A. doi:10.1126/science.235.4793.1196. PMID 17818979.

67. Bickers, N.E.; Scalapino, D. J.; Scalettar, R. T. (1987). "CDW and SDW mediated pairing interactions". Int. J. Mod. Phys. B. 1 (3n04): 687–695. Bibcode:1987IJMPB...1..687B. doi:10.1142/S0217979287001079.

68. Inui, Masahiko; Doniach, Sebastian; Hirschfeld, Peter J.; Ruckenstein, Andrei E.; Zhao, Z.; Yang, Q.; Ni, Y.; Liu, G. (1988). "Coexistence of antiferromagnetism and superconductivity in a mean-field theory of high-Tc superconductors". Phys. Rev. B. 37 (10): 5182–5185. Bibcode:1988PhRvB..37.5182D. doi:10.1103/PhysRevB.37.5182.

69. Gros, Claudius; Poilblanc, Didier; Rice, T. Maurice; Zhang, F. C. (1988). "Superconductivity in correlated wavefunctions". Physica C. 153–155: 543–548. Bibcode:1988PhyC..153..543G. doi:10.1016/0921-4534(88)90715-0.

70. Kotliar, Gabriel; Liu, Jialin (1988). "Superexchange mechanism and d-wave superconductivity". Physical. Review B. 38 (7): 5182. Bibcode:1988PhRvB..38.5142K. doi:10.1103/PhysRevB.38.5142.

71. Schilling, A.; Cantoni, M.; Guo, J. D.; Ott, H. R. (1993). "Superconductivity in the Hg-Ba-Ca-Cu-O system". Nature. 363 (6424): 56–58. Bibcode:1993Natur.363...56S. doi:10.1038/363056a0.

72. Mourachkine, A. (2004). Room-Temperature Superconductivity. Cambridge International Science Publishing. arXiv:cond-mat/0606187 . ISBN 1-904602-27-4.

73. Piazza, B. Dalla (15 December 2014). "Fractional excitations in the square-lattice quantum antiferromagnet". Nature Physics 11, 62–68. doi:10.1038/nphys3172. Retrieved 23 December 2014.

74. "How electrons split: New evidence of exotic behaviors". Nanowerk. École Polytechnique Fédérale de Lausanne. Dec 23, 2014. Retrieved Dec 23, 2014.

75. Hazen, R.; Finger, L.; Angel, R.; Prewitt, C.; Ross, N.; Mao, H.; Hadidiacos, C.; Hor, P.; Meng, R.; Chu, C. (1987). "Crystallographic description of phases in the Y-Ba-Cu-O superconductor". Physical Review B. 35 (13): 7238–7241. Bibcode:1987PhRvB..35.7238H. doi:10.1103/PhysRevB.35.7238.

76. Khare, Neeraj (2003). Handbook of High-Temperature Superconductor Electronics. CRC Press. ISBN 0-8247-0823-7.

77. Hermann, Allen M. and Yakhmi, J.V. eds. (1994) Thallium-Based High-Temperature Superconductors, Marcel Dekker

78. Hazen, R.; Prewitt, C.; Angel, R.; Ross, N.; Finger, L.; Hadidiacos, C.; Veblen,

D.; Heaney, P.; Hor, P.; Meng, R.; Sun, Y.; Wang, Y.; Xue, Y.; Huang, Z.; Gao, L.; Bechtold, J.; Chu, C. (1988). "Superconductivity in the high-Tc Bi-Ca-Sr-Cu-O system: Phase identification". Physical Review Letters. 60 (12): 1174–1177. Bibcode:1988PhRvL..60.1174H. doi:10.1103/PhysRevLett.60.1174.

79. Tarascon, J.; McKinnon, W.; Barboux, P.; Hwang, D.; Bagley, B.; Greene, L.; Hull, G.; Lepage, Y.; Stoffel, N.; Giroud, M. (1988). "Preparation, structure, and properties of the superconducting compound series Bi2Sr2Can−1CunOy with n = 1, 2, and 3". Physical Review B. 38 (13): 8885–8892. Bibcode:1988PhRvB..38.8885T. doi:10.1103/PhysRevB.38.8885.

80. Sheng, Z. Z.; Hermann, A. M.; El Ali, A; Almasan, C; Estrada, J; Datta, T; Matson, R. J. (1988). "Superconductivity at 90 K in the Tl-Ba-Cu-O system". Physical Review Letters. 60 (10): 937–940. Bibcode:1988PhRvL..60..937S. doi:10.1103/PhysRevLett.60.937. PMID 10037895.

81. Sheng, Z. Z.; Hermann, A. M. (1988). "Superconductivity in the rare-earth-free Tl-Ba-Cu-O system above liquid-nitrogen temperature". Nature. 332 (6159): 55–58. Bibcode:1988Natur.332...55S. doi:10.1038/332055a0.

82. Putilin, S. N.; Antipov, E. V.; Chmaissem, O.; Marezio, M. (1993). "Superconductivity at 94 K in HgBa2Cu04+δ". Nature. 362 (6417): 226–228. Bibcode:1993Natur.362..226P. doi:10.1038/362226a0.

83. Schilling, A.; Cantoni, M.; Guo, J. D.; Ott, H. R. (1993). "Superconductivity above 130 K in the Hg–Ba–Ca–Cu–O system". Nature. 363 (6424): 56–58. Bibcode:1993Natur.363...56S. doi:10.1038/363056a0.

84. Chu, C. W.; Gao, L.; Chen, F.; Huang, Z. J.; Meng, R. L.; Xue, Y. Y. (1993). "Superconductivity above 150 K in HgBa2Ca2Cu3O8+δ at high pressures". Nature. 365 (6444): 323–325. Bibcode:1993Natur.365..323C. doi:10.1038/365323a0.

85. Shi, Donglu; Boley, Mark S.; Chen, J. G.; Xu, Ming; Vandervoort, K.; Liao, Y. X.; Zangvil, A.; Akujieze, Justin; Segre, Carlo (1989). "Origin of enhanced growth of the 110 K superconducting phase by Pb doping in the Bi-Sr-Ca-Cu-O system". Applied Physics Letters. 55 (7): 699. Bibcode:1989ApPhL..55..699S. doi:10.1063/1.101573.

86. Norman, Michael R. (2008). "Trend: High-temperature superconductivity in the iron pnictides". Physics. 1 (21): 21. Bibcode:2008PhyOJ...1...21N. doi:10.1103/Physics.1.21. Retrieved 30 March 2012.

87. "High-Temperature Superconductivity: The Cuprates". Devereaux group. Stanford University. Retrieved 30 March 2012.

88. http://www.nature.com/nphys/journal/v6/n8/full/nphys1687.html

89. M. Buchanan (2001). "Mind the pseudogap". Nature. 409 (6816): 8–11. doi:10.1038/35051238. PMID 11343081.

90. Sanna, S.; Allodi, G.; Concas, G.; Hillier, A.; Renzi, R. (2004). "Nanoscopic Coexistence of Magnetism and Superconductivity in YBa2Cu3O6+x Detected by Muon Spin Rotation". Physical Review Letters. 93 (20): 207001. arXiv:cond-mat/0403608 . Bibcode:2004PhRvL..93t7001S. doi:10.1103/PhysRevLett.93.207001.

91. C. Hartinger. "DFG FG 538 – Doping Dependence of Phase transitions and Ordering Phenomena in Cuprate Superconductors". Wmi.badw-muenchen.de. Retrieved 2009-10-29.

92. Luetkens, H; Klauss, H. H.; Kraken, M; Litterst, F. J.; Dellmann, T; Klingeler, R; Hess, C; Khasanov, R; Amato, A; Baines, C; Kosmala, M; Schumann, O. J.; Braden, M; Hamann-Borrero, J; Leps, N; Kondrat, A; Behr, G; Werner, J; Büchner, B (2009). "Electronic phase diagram of the LaO1-xFxFeAs superconductor". Nature Materials. 8 (4): 305–9. arXiv:0806.3533 . Bibcode:2009NatMa...8..305L. doi:10.1038/nmat2397. PMID 19234445.

93. Drew, A. J.; Niedermayer, Ch; Baker, P. J.; Pratt, F. L.; Blundell, S. J.; Lancaster, T; Liu, R. H.; Wu, G; Chen, X. H.; Watanabe, I; Malik, V. K.; Dubroka, A; Rössle, M; Kim, K. W.; Baines, C; Bernhard, C (2009). "Coexistence of static magnetism and superconductivity in SmFeAsO1-xFx as revealed by muon spin rotation". Nature Materials. 8 (4): 310–314. arXiv:0807.4876 . Bibcode:2009NatMa...8..310D. doi:10.1038/nmat2396. PMID 19234446.

94. Sanna, S.; De Renzi, R.; Lamura, G.; Ferdeghini, C.; Palenzona, A.; Putti, M.; Tropeano, M.; Shiroka, T. (2009). "Competition between magnetism and superconductivity at the phase boundary of doped SmFeAsO pnictides". Physical Review B. 80 (5): 052503. arXiv:0902.2156 . Bibcode:2009PhRvB..80e2503S. doi:10.1103/PhysRevB.80.052503.

95. Zhao, J; Huang, Q; de la Cruz, C; Li, S; Lynn, J. W.; Chen, Y; Green, M. A.; Chen, G. F.; Li, G; Li, Z; Luo, J. L.; Wang, N. L.; Dai, P (2008). "Structural and magnetic phase diagram of CeFeAsO1-xFx and its relation to high-temperature superconductivity". Nature Materials. 7 (12): 953–959. arXiv:0806.2528 . Bibcode:2008NatMa...7..953Z. doi:10.1038/nmat2315. PMID 18953342.

96. Chu, Jiun-Haw; Analytis, James; Kucharczyk, Chris; Fisher, Ian (2008). "Determination of the phase diagram of the electron doped superconductor Ba(Fe1-xCox)2As2". Physical Review B. 79: 014506. arXiv:0811.2463 . Bibcode:2009PhRvB..79a4506C. doi:10.1103/PhysRevB.79.014506.

97. Kamihara, Y; Hiramatsu, H; Hirano, M; Kawamura, R; Yanagi, H; Kamiya, T; Hosono, H (2006). "Iron-Based Layered Superconductor: LaOFeP". Journal of the American Chemical Society. 128 (31): 10012–10013. doi:10.1021/ja063355c. PMID 16881620.

98. Kamihara, Y; Watanabe, T; Hirano, M; Hosono, H (2008). "Iron-Based Layered Superconductor La[O1-xFx]FeAs (x = 0.05–0.12) with Tc = 26 K". Journal of the American Chemical Society. 130 (11): 3296–3297. doi:10.1021/ja800073m. PMID 18293989.

99. Takahashi, H; Igawa, K; Arii, K; Kamihara, Y; Hirano, M; Hosono, H (2008). "Superconductivity at 43 K in an iron-based layered compound LaO1-xFxFeAs". Nature. 453 (7193): 376–378. Bibcode:2008Natur.453..376T. doi:10.1038/nature06972. PMID 18432191.

100. Wang, Qing-Yan; Li, Zhi; Zhang, Wen-Hao; Zhang, Zuo-Cheng; Zhang, Jin-Song; Li, Wei; Ding, Hao; Ou, Yun-Bo; Deng, Peng; Chang, Kai; Wen, Jing;

Song, Can-Li; He, Ke; Jia, Jin-Feng; Ji, Shuai-Hua; Wang, Ya-Yu; Wang, Li-Li; Chen, Xi; Ma, Xu-Cun; Xue, Qi-Kun (2012). "Interface-Induced High-Temperature Superconductivity in Single Unit-Cell FeSe Films on SrTiO3". Chin. Phys. Lett. 29 (3): 037402. arXiv:1201.5694 . Bibcode:2012ChPhL..29c7402W. doi:10.1088/0256-307X/29/3/037402.

101. Liu, Defa; Zhang, Wenhao; Mou, Daixiang; He, Junfeng; Ou, Yun-Bo; Wang, Qing-Yan; Li, Zhi; Wang, Lili; Zhao, Lin; He, Shaolong; Peng, Yingying; Liu, Xu; Chen, Chaoyu; Yu, Li; Liu, Guodong; Dong, Xiaoli; Zhang, Jun; Chen, Chuangtian; Xu, Zuyan; Hu, Jiangping; Chen, Xi; Ma, Xucun; Xue, Qikun; Zhou, X.J. (2012). "Electronic origin of high-temperature superconductivity in single-layer FeSe superconductor". Nat. Comms. 3: 931. arXiv:1202.5849 . Bibcode:2012NatCo...3E.931L. doi:10.1038/ncomms1946.

102. He, Shaolong; He, Junfeng; Zhang, Wenhao; Zhao, Lin; Liu, Defa; Liu, Xu; Mou, Daixiang; Ou, Yun-Bo; Wang, Qing-Yan; Li, Zhi; Wang, Lili; Peng, Yingying; Liu, Yan; Chen, Chaoyu; Yu, Li; Liu, Guodong; Dong, Xiaoli; Zhang, Jun; Chen, Chuangtian; Xu, Zuyan; Chen, Xi; Ma, Xucun; Xue, Qikun; Zhou, X. J. (2013). "Phase diagram and electronic indication of high-temperature superconductivity at 65 K in single-layer FeSe films". Nat. Mat. 12 (7): 605–610. arXiv:1207.6823 . Bibcode:2013NatMa..12..605H. doi:10.1038/NMAT3648.

103. Jian-Feng Ge; et al. (2014). "Superconductivity in single-layer films of FeSe with a transition temperature above 100 K". Nature Materials. 1406: 3435. arXiv:1406.3435 . Bibcode:2015NatMa..14..285G. doi:10.1038/nmat4153.

104. Wu, G; Xie, Y L; Chen, H; Zhong, M; Liu, R H; Shi, B C; Li, Q J; Wang, X F; Wu, T; Yan, Y J; Ying, J J; Chen, X H (2009). "Superconductivity at 56 K in Samarium-doped SrFeAsF". Journal of Physics: Condensed Matter. 21 (3): 142203. arXiv:0811.0761 . Bibcode:2009JPCM...21n2203W. doi:10.1088/0953-8984/21/14/142203.

105. Rotter, M; Tegel, M; Johrendt, D (2008). "Superconductivity at 38 K in the Iron Arsenide (Ba1-xKx)Fe2As2". Physical Review Letters. 101 (10): 107006. arXiv:0805.4630 . Bibcode:2008PhRvL.101j7006R. doi:10.1103/PhysRevLett.101.107006. PMID 18851249.

106. Sasmal, K; Lv, B; Lorenz, B; Guloy, A. M.; Chen, F; Xue, Y. Y.; Chu, C. W. (2008). "Superconducting Fe-Based Compounds (A1-xSrx)Fe2As2with A=K and Cs with Transition Temperatures up to 37 K". Physical Review Letters. 101 (10): 107007. Bibcode:2008PhRvL.101j7007S. doi:10.1103/PhysRevLett.101.107007. PMID 18851250.

107. Pitcher, M. J.; Parker, D. R.; Adamson, P; Herkelrath, S. J.; Boothroyd, A. T.; Ibberson, R. M.; Brunelli, M; Clarke, S. J. (2008). "Structure and superconductivity of LiFeAs". Chemical Communications. 2008 (45): 5918–5920. doi:10.1039/b813153h. PMID 19030538.

108. Tapp, Joshua H.; Tang, Zhongjia; Lv, Bing; Sasmal, Kalyan; Lorenz, Bernd; Chu, Paul C. W.; Guloy, Arnold M. (2008). "LiFeAs: An intrinsic FeAs-based superconductor with Tc=18 K". Physical Review B. 78 (6): 060505. arXiv:0807.2274 . Bibcode:2008PhRvB..78f0505T. doi:10.1103/PhysRevB.78.060505.

109. Parker, D. R.; Pitcher, M. J.; Baker, P. J.; Franke, I; Lancaster, T; Blundell, S. J.; Clarke, S. J. (2009). "Structure, antiferromagnetism and superconductivity of the layered iron arsenide NaFeAs". Chemical Communications (16): 2189–2191. doi:10.1039/b818911k. PMID 19360189.

110. Hsu, F. C.; Luo, J. Y.; Yeh, K. W.; Chen, T. K.; Huang, T. W.; Wu, P. M.; Lee, Y. C.; Huang, Y. L.; Chu, Y. Y.; Yan, D. C.; Wu, M. K. (2008). "Superconductivity in the PbO-type structure α-FeSe". Proceedings of the National Academy of Sciences. 105 (38): 14262–14264. Bibcode:2008PNAS..10514262H. doi:10.1073/pnas.0807325105. PMC 2531064 . PMID 18776050.

111. Kordyuk, A. A. (2012). "Iron-based superconductors: Magnetism, superconductivity, and electronic structure (Review Article)" (PDF). Low Temp. Phys. 38 (9): 888–899. arXiv:1209.0140 . Bibcode:2012LTP....38..888K. doi:10.1063/1.4752092.

112. Lee, Chul-Ho; Iyo, Akira; Eisaki, Hiroshi; Kito, Hijiri; Teresa Fernandez-Diaz, Maria; Ito, Toshimitsu; Kihou, Kunihiro; Matsuhata, Hirofumi; Braden, Markus; Yamada, Kazuyoshi (2008). "Effect of Structural Parameters on Superconductivity in Fluorine-Free LnFeAsO1-y (Ln = La, Nd)". Journal of the Physical Society of Japan. 77 (8): 083704. arXiv:0806.3821 . Bibcode:2008JPSJ...77h3704L. doi:10.1143/JPSJ.77.083704.

113. Drozdov, A.; Eremets, M. I.; Troyan, I. A. (2014). "Conventional superconductivity at 190 K at high pressures". arXiv:1412.0460 .

114. Ge, Y. F.; Zhang, F.; Yao, Y. G. (2016). "First-principles demonstration of superconductivity at 280 K in hydrogen sulfide with low phosphorus substitution". Phys. Rev. B. 93 (22): 224513. doi:10.1103/PhysRevB.93.224513.

115. Preuss, Paul. "A Most Unusual Superconductor and How It Works". Berkeley Lab. Retrieved 12 March 2012.

116. Hebard, A. F.; Rosseinsky, M. J.; Haddon, R. C.; Murphy, D. W.; Glarum, S. H.; Palstra, T. T. M.; Ramirez, A. P.; Kortan, A. R. (1991). "Superconductivity at 18 K in potassium-doped C60". Nature. 350 (6319): 600–601. Bibcode:1991Natur.350..600H. doi:10.1038/350600a0.

117. Ganin, A. Y.; Takabayashi, Y; Khimyak, Y. Z.; Margadonna, S; Tamai, A; Rosseinsky, M. J.; Prassides, K (2008). "Bulk superconductivity at 38 K in a molecular system". Nature Materials. 7 (5): 367–71. Bibcode:2008NatMa...7..367G. doi:10.1038/nmat2179. PMID 18425134.

118. Ashcroft, N. W. (1968). "Metallic Hydrogen: A High-Temperature Superconductor?". Physical Review Letters. 21 (26): 1748–1749. Bibcode:1968PhRvL..21.1748A. doi:10.1103/PhysRevLett.21.1748.

119. Monthoux, P.; Balatsky, A.; Pines, D. (1992). "Weak-coupling theory of high-temperature superconductivity in the antiferromagnetically correlated copper oxides". Physical Review B. 46 (22): 14803–14817. Bibcode:1992PhRvB..4614803M. doi:10.1103/PhysRevB.46.14803.

120. Chakravarty, S; Sudbø, A; Anderson, P. W.; Strong, S (1993). "Interlayer Tunneling and Gap Anisotropy in High-Temperature Superconductors". Science. 261 (5119):

337–340. Bibcode:1993Sci...261..337C. doi:10.1126/science.261.5119.337. PMID 17836845.

121. Geshkenbein, V.; Larkin, A.; Barone, A. (1987). "Vortices with half magnetic flux quanta in heavy-fermion superconductors". Physical Review B. 36 (1): 235–238. Bibcode:1987PhRvB..36..235G. doi:10.1103/PhysRevB.36.235. PMID 9942041.

122. Kirtley, J. R.; Tsuei, C. C.; Sun, J. Z.; Chi, C. C.; Yu-Jahnes, Lock See; Gupta, A.; Rupp, M.; Ketchen, M. B. (1995). "Symmetry of the order parameter in the high-Tc superconductor YBa2Cu3O7-δ". Nature. 373 (6511): 225–228. Bibcode:1995Natur.373..225K. doi:10.1038/373225a0.

123. Kirtley, J. R.; Tsuei, C. C.; Ariando, A.; Verwijs, C. J. M.; Harkema, S.; Hilgenkamp, H. (2006). "Angle-resolved phase-sensitive determination of the in-plane gap symmetry in YBa2Cu3O7-δ". Nature Physics. 2 (3): 190–194. Bibcode:2006NatPh...2..190K. doi:10.1038/nphys215.

124. Tsuei, C. C.; Kirtley, J. R.; Ren, Z. F.; Wang, J. H.; Raffy, H.; Li, Z. Z. (1997). "Pure dx2-y2 order-parameter symmetry in the tetragonal superconductor Tl2Ba2CuO6+δ". Nature. 387 (6632): 481–483. Bibcode:1997Natur.387..481T. doi:10.1038/387481a0.

CHAPTER 7

Alloy: Structure and Properties

Chapter Contents

- **Introduction**
- **Terminology**
- **Theory**
- **Heat-Treatable Alloys**
- **Substitutional and Interstitial Alloys**
- **History and Examples**

Low alloy steel is a type of steel that has other materials added to it, but the other materials typically make up a small amount of the entire steel. Steel commonly is an alloy consisting of carbon and iron, but low alloy steel often adds hard metals such as nickel and chromium. This steel usually is hard and difficult to weld, but the steel's characteristics can be changed depending on the inclusions. This most commonly is used to make large pipes for the oil industry, but it also can be used for construction and military purposes.

INTRODUCTION

An **alloy** is a mixture of metals or a mixture of a metal and another element. Alloys are defined by a metallic bonding character. An alloy may be a solid solution of metal elements (a single phase) or a mixture of metallic phases (two or more solutions). Intermetallic compounds are alloys with a defined stoichiometry and crystal structure. Zintl phases are also sometimes considered alloys depending on bond types (see also: Van Arkel-Ketelaar triangle for information on classifying bonding in binary compounds).

Alloys are used in a wide variety of applications. In some cases, a combination of metals may reduce the overall cost of the material while preserving important properties. In other cases, the combination of metals imparts synergistic properties to the constituent metal elements such as corrosion resistance or mechanical strength. Examples of alloys are steel, solder, brass, pewter, duralumin, bronze and amalgams.

The alloy constituents are usually measured by mass. Alloys are usually classified as substitutional or interstitial alloys, depending on the atomic arrangement that forms the alloy. They can be further classified as homogeneous

Liquid bronze, being poured into molds during casting.

A brass lamp.

An alloy is a mixture of chemical elements, which forms an impure substance (admixture) that retains the characteristics of a metal. An alloy is distinct from an impure metal in that, with an alloy, the added elements are well controlled to produce desirable properties, while impure metals such as wrought iron, are less controlled, but are often considered useful. Alloys are made by mixing two or more elements, at least one of which is a metal. This is usually called the primary metal or the base metal, and the name of this metal may also be the name of the alloy. The other constituents may or may not be metals but, when mixed with the molten base, they will be soluble and dissolve into the mixture.

The mechanical properties of alloys will often be quite different from those of its individual constituents. A metal that is normally very soft (malleable), such as aluminium, can be altered by alloying it with another soft metal, such as copper. Although both metals are very soft and ductile, the resulting aluminium alloy will have much greater strength. Adding a small amount of non-metallic carbon to iron trades its great ductility for the greater strength of an alloy called steel. Due to its very-high strength, but still substantial toughness, and its ability to be greatly altered by heat treatment, steel is one of the most useful and common alloys in modern use. By adding chromium to steel, its resistance to corrosion can be enhanced, creating stainless steel, while adding silicon will alter its electrical characteristics, producing silicon steel.

Although the elements of an alloy usually must be soluble in the liquid state, they may not always be soluble in the solid state. If the metals remain soluble when solid, the alloy forms a solid solution, becoming a homogeneous structure consisting of identical crystals, called a phase. If as the mixture cools the constituents become insoluble, they may separate to form two or more different types of crystals, creating a heterogeneous microstructure of different phases, some with more of one constituent than the other phase has. However, in other alloys, the insoluble elements may not separate until after crystallization occurs. These alloys are called intermetallic alloys because, if cooled very quickly, they first crystallize as a homogeneous phase, but they are supersaturated and unstable with the secondary constituents. As time passes, the atoms of these supersaturated alloys separate from the crystal lattice, becoming more stable, and form intermetallic (within the crystal lattice) phases that serve to reinforce the crystals internally.

Some alloys, such as electrum which is an alloy consisting of silver and gold, occur naturally. Meteorites are sometimes made of naturally occurring alloys of iron and nickel, but are not native to the Earth. One of the first alloys made by humans was bronze, which is a mixture of the metals tin and copper. Bronze was an extremely useful alloy to the ancients, because it is much stronger and harder than either of its components. Steel was another common alloy. However, in ancient times, it could only be created as an accidental byproduct from the heating of iron ore in fires (smelting) during the manufacture of iron. Other ancient alloys include pewter, brass and pig iron. In the modern age, steel can be created in many forms. Carbon steel can be made by varying only the carbon content, producing soft alloys like mild steel or hard alloys like spring steel. Alloy steels can be made by adding other elements, such as chromium, molybdenum, vanadium or nickel, resulting in alloys such as high-speed steel or tool steel. Small amounts of manganese are usually alloyed with most modern steels because of its ability to remove unwanted impurities, like phosphorus, sulfur and oxygen, which can have detrimental effects on the alloy. However, most alloys were not created until the 1900s, such as various aluminium, titanium, nickel, and magnesium alloys. Some modern superalloys, such as incoloy, inconel, and hastelloy, may consist of a multitude of different elements.

TERMINOLOGY

A gate valve, made from Inconel.

The term alloy is used to describe a mixture of atoms in which the primary constituent is a metal. The primary metal is called the *base*, the *matrix*, or the *solvent*. The secondary constituents are often called *solutes*. If there is a mixture of only two types of atoms (not counting impurities) such as a copper-nickel alloy, then it is called a *binary alloy.* If there are three types of atoms forming the mixture, such as iron, nickel and chromium, then it is called a *ternary alloy.* An alloy with four constituents is a *quaternary alloy,* while a five-part alloy is termed a *quinary alloy.* Because the percentage of each constituent can be varied, with any mixture the entire range of possible variations is called a *system*. In this respect, all of the various forms of an alloy containing only two constituents, like iron and carbon, is called a *binary system,* while all of the alloy combinations possible with a ternary alloy, such as alloys of iron, carbon and chromium, is called a *ternary system*.

Although an alloy is technically an impure metal, when referring to alloys, the term "impurities" usually denotes those elements which are not desired. Such impurities are introduced from the base metals and alloying elements, but are removed during processing. For instance, sulfur is a common impurity in steel. Sulfur combines readily with iron to form iron sulfide, which is very brittle, creating weak spots in the steel. Lithium, sodium and calcium are common impurities in aluminium alloys, which can have adverse effects on the structural integrity of castings. Conversely, otherwise pure-metals that simply contain unwanted impurities are often called "impure metals" and are not usually referred to as alloys. Oxygen, present in the air, readily combines with most metals to form metal oxides; especially at higher temperatures encountered during alloying. Great care is often taken during the alloying process to remove excess impurities, using fluxes, chemical additives, or other methods of extractive metallurgy.

In practice, some alloys are used so predominantly with respect to their base metals that the name of the primary constituent is also used as the name of the alloy. For example, 14 karat gold is an alloy of gold with other elements. Similarly, the silver used in jewelry and the aluminium used as a structural building material are also alloys.

The term «alloy» is sometimes used in everyday speech as a synonym for a particular alloy. For example, automobile wheels made of an aluminium alloy are commonly referred to as simply "alloy wheels», although in point of fact steels and most other metals in practical use are also alloys. Steel is such a common alloy that many items made from it, like wheels, barrels, or girders, are simply referred to by the name of the item, assuming it is made of steel. When made from other materials, they are typically specified as such, (i.e.: «bronze wheel», «plastic barrel», or «wood girder»).

THEORY

Alloying a metal is done by combining it with one or more other elements that often enhance its properties. For example, the combination of carbon with iron produces steel, which is stronger than iron, its primary element. The electrical and thermal conductivity of alloys is usually lower than that of the pure metals. The physical properties, such as density, reactivity, Young›s modulus of an alloy may not differ greatly from those of its base element, but engineering properties such as tensile strength, ductility, and shear strength may be substantially different from those of the constituent materials. This is sometimes a result of the sizes of the atoms in the alloy, because larger atoms exert a compressive force on neighboring atoms, and smaller atoms exert a tensile force on their neighbors, helping the alloy resist deformation. Sometimes alloys may exhibit marked differences in behavior even when small amounts of one element are present. For example, impurities in semiconducting ferromagnetic alloys lead to different properties, as first predicted by White, Hogan, Suhl, Tian Abrie and Nakamura. Some alloys are made by melting and mixing two or more metals. Bronze, an alloy of copper and tin, was the first alloy discovered, during the prehistoric period now known as the bronze age. It was harder than pure copper and originally used to make tools and weapons, but was later superseded by metals and alloys with better properties. In later times bronze has been used for ornaments, bells, statues, and bearings. Brass is an alloy made from copper and zinc.

Unlike pure metals, most alloys do not have a single melting point, but a melting range during which the material is a mixture of solid and liquid phases (a slush). The temperature at which melting begins is called the solidus, and the temperature when melting is just complete is called the liquidus. For many alloys there is a particular alloy proportion (in some cases more than one), called either a eutectic mixture or a peritectic composition, which gives the alloy a unique and low melting point, and no liquid/solid slush transition.

HEAT-TREATABLE ALLOYS

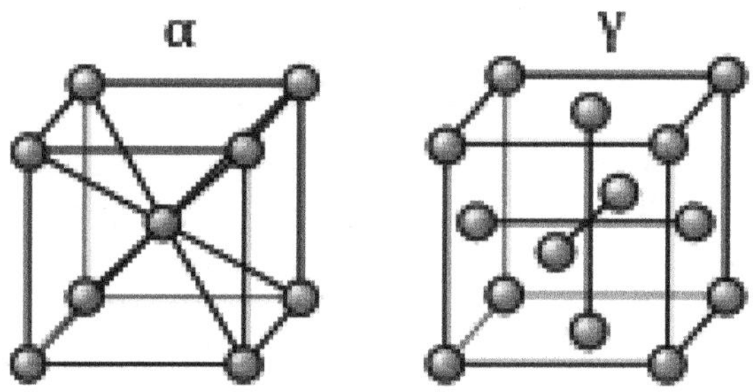

Allotropes of iron, (alpha iron and gamma iron) showing the differences in atomic arrangement.

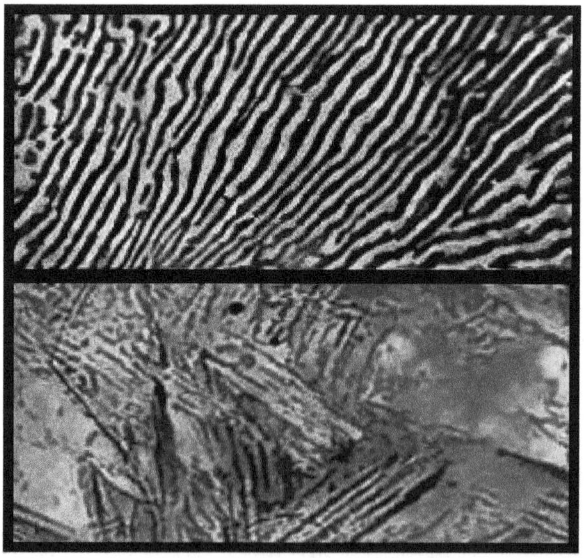

Photomicrographs of steel. Top photo: Annealed (slowly cooled) steel forms a heterogeneous, lamellar microstructure called pearlite, consisting of the phases cementite (light) and ferrite (dark). Bottom photo: Quenched (quickly cooled) steel forms a single phase called martensite, in which the carbon remains trapped within the crystals, creating internal stresses.

Alloying elements are added to a base metal, to induce hardness, toughness, ductility, or other desired properties. Most metals and alloys can be work hardened by creating defects in their crystal structure. These defects are created during plastic deformation by hammering, bending, extruding, etcetera, and are permanent unless the metal is recrystallized. Otherwise, some alloys can also have their properties altered by heat treatment. Nearly all metals can be softened by annealing, which recrystallizes the alloy and repairs the defects, but not as many can be hardened by controlled heating and cooling. Many alloys of aluminium, copper, magnesium, titanium, and nickel can be strengthened to some degree by some method of heat treatment, but few respond to this to the same degree as does steel.

The base metal iron of the iron-carbon alloy known as steel, undergoes a change in the arrangement (allotropy) of the atoms of its crystal matrix at a certain temperature (usually between 1,500 °F (820 °C) and 1,600 °F (870 °C), depending on carbon content). This allows the smaller carbon atoms to enter the interstices of the iron crystal. When this diffusion happens, the carbon atoms are said to be in *solution* in the iron, forming a particular single, homogeneous, crystalline phase called austenite. If the steel is cooled slowly, the carbon can diffuse out of the iron and it will gradually revert to its low temperature allotrope. During slow cooling, the carbon atoms will no longer be as soluble with the iron, and will be forced to precipitate out of solution, nucleating into a more concentrated form of iron carbide (Fe_3C) in the spaces between the pure iron crystals. The steel then becomes heterogeneous, as it is formed of two phases, the iron-carbon phase called cementite (or carbide), and pure iron ferrite. Such a heat treatment produces a steel that is rather soft. If the steel is cooled quickly, however, the carbon atoms will not have time to diffuse and precipitate out as carbide, but will be trapped within the iron crystals. When rapidly cooled, a diffusionless (martensite) transformation occurs, in which the carbon atoms become trapped in solution. This causes the iron crystals to deform as the crystal structure tries to change to its low temperature state, leaving those crystals very hard but much less ductile (brittle).

While the high strength of steel results when diffusion and precipitation is prevented (forming martinsite), most heat-treatable alloys are precipitation hardening alloys, that depend on the diffusion of alloying elements to achieve their strength. When heated to form a solution and then cooled quickly, these alloys become much softer than normal, during the diffusionless transformation, but then harden as they age. The solutes in these alloys will precipitate over time, forming intermetallic phases, which are difficult to discern from the base metal. Unlike steel, in which the solid solution separates into different crystal phases (carbide and ferrite), precipitation hardening alloys form different phases within the same crystal. These intermetallic alloys appear homogeneous in crystal structure, but tend to behave heterogeneously, becoming hard and somewhat brittle.

SUBSTITUTIONAL AND INTERSTITIAL ALLOYS

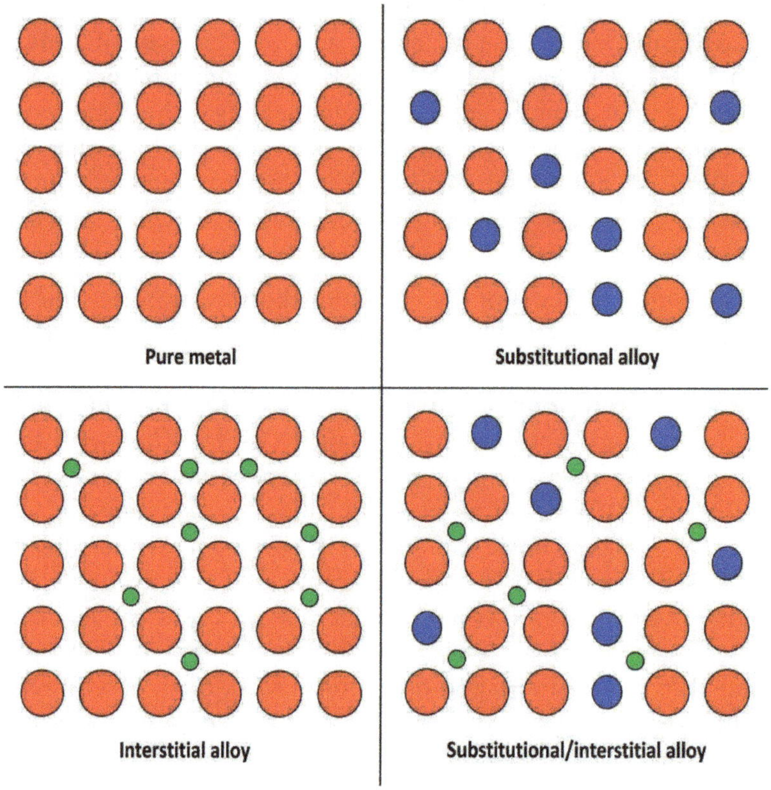

Different atomic mechanisms of alloy formation, showing pure metal, substitutional, interstitial, and a combination of the two.

When a molten metal is mixed with another substance, there are two mechanisms that can cause an alloy to form, called *atom exchange* and the *interstitial mechanism*. The relative size of each element in the mix plays a primary role in determining which mechanism will occur. When the atoms are relatively similar in size, the atom exchange method usually happens, where some of the atoms composing the metallic crystals are substituted with atoms of the other constituent. This is called a *substitutional alloy*. Examples of substitutional alloys include bronze and brass, in which some of the copper atoms are substituted with either tin or zinc atoms respectively. In the case of the interstitial mechanism, one atom is usually much smaller than the other and can not successfully substitute for the other type of atom in the crystals of the base metal. Instead, the smaller atoms become trapped in the spaces between the atoms of the crystal matrix, called the *interstices*. This is referred to as an *interstitial alloy*. Steel is an example of an interstitial alloy, because the very small carbon atoms fit into interstices of the iron matrix. Stainless steel is an example of a combination of interstitial and substitutional alloys, because the carbon atoms fit into the interstices, but some of the iron atoms are substituted by nickel and chromium atoms.

HISTORY AND EXAMPLES

Meteoric Iron

A meteorite and a hatchet that was forged from meteoric iron.

The use of alloys by humans started with the use of meteoric iron, a naturally occurring alloy of nickel and iron. It is the main constituent of iron meteorites which occasionally fall down on Earth from outer space. As no metallurgic processes were used to separate iron from nickel, the alloy was used as it was. Meteoric iron could be forged from a red heat to make objects such as tools, weapons, and nails. In many cultures it was shaped by cold hammering into knives and arrowheads. They were often used as anvils. Meteoric iron was very rare and valuable, and difficult for ancient people to work.

Bronze and Brass

Bronze axe 1100 BC

Bronze doorknocker.

Iron is usually found as iron ore on Earth, except for one deposit of native iron in Greenland, which was used by the Inuit people. Native copper, however, was found worldwide, along with silver, gold and platinum, which were also used to make tools, jewelry, and other objects since Neolithic times. Copper was the hardest of these metals, and the most widely distributed. It became one of the most important metals to the ancients. Eventually, humans learned to smelt metals such as copper and tin from ore, and, around 2500 BC, began alloying the two metals to form bronze, which is much harder than its ingredients. Tin was rare, however, being found mostly in Great Britain. In the Middle East, people began alloying copper with zinc to form brass. Ancient civilizations took into account the mixture and the various properties it produced, such as hardness, toughness and melting point, under various conditions of temperature and work hardening, developing much of the information contained in modern alloy phase diagrams. Arrowheads from the Chinese Qin dynasty (around 200 BC) were often constructed with a hard bronze-head, but a softer bronze-tang, combining the alloys to prevent both dulling and breaking during use.

Amalgams

Mercury has been smelted from cinnabar for thousands of years. Mercury dissolves many metals, such as gold, silver, and tin, to form amalgams (an alloy in a soft paste, or liquid form at ambient temperature). Amalgams have been used since 200 BC in China for plating objects with precious metals, called gilding, such as armor and mirrors. The ancient Romans often used mercury-tin amalgams for gilding their armor. The amalgam was applied as a paste and then heated until the mercury vaporized, leaving the gold, silver, or tin behind. Mercury was often used in mining, to extract precious metals like gold and silver from their ores.

Precious-metal Alloys

Electrum, a natural alloy of silver and gold, was often used for making coins.

Many ancient civilizations alloyed metals for purely aesthetic purposes. In ancient Egypt and Mycenae, gold was often alloyed with copper to produce red-gold, or iron to produce a bright burgundy-gold. Gold was often found alloyed with silver or other metals to produce various types of colored gold. These metals were also used to strengthen each other, for more practical purposes. Copper was often added to silver to make sterling silver, increasing its strength for use in dishes, silverware, and other practical items. Quite often, precious metals were alloyed with less valuable substances as a means to deceive buyers.Around 250 BC, Archimedes was commissioned by the king to find a way to check the purity of the gold in a crown, leading to the famous bath-house shouting of "Eureka!" upon the discovery of Archimedes› principle.

Pewter

The term pewter covers a variety of alloys consisting primarily of tin. As a pure metal, tin was much too soft to be used for any practical purpose. However, in the Bronze age, tin was a rare metal and, in many parts of Europe and the Mediterranean, was often valued higher than gold. To make jewelry, forks and spoons, or other objects from tin, it was usually alloyed with other metals to increase its strength and hardness. These metals were typically lead, antimony, bismuth or copper. These solutes sometimes were added individually in varying amounts, or added together, making a wide variety of things, ranging from practical items, like dishes, surgical tools, candlesticks or funnels, to decorative items such as ear rings and hair clips.

The earliest examples of pewter come from ancient Egypt, around 1450 BC. The use of pewter was widespread across Europe, from France to Norway and Britain (where most of the ancient tin was mined) to the Near East. The alloy was also used in China and the Far East, arriving in Japan around 800 AD, where it was used for making objects like ceremonial vessels, tea canisters, or chalices used in shinto shrines.

Steel and Pig Iron

The first known smelting of iron began in Anatolia, around 1800 BC. Called the bloomery process, it produced very soft but ductile wrought iron. By 800 BC, iron-making technology had spread to Europe, arriving in Japan around 700 AD. Pig iron, a very hard but brittle alloy of iron and carbon, was being produced in China as early as 1200 BC, but did not arrive in Europe until the Middle Ages. Pig iron has a lower melting point than iron, and was used for making cast-iron. However, these metals found little practical use until the introduction of crucible steel around 300 BC. These steels were of poor quality, and the introduction of pattern welding, around the 1st century AD, sought to balance the extreme properties of the alloys by laminating them, to create a tougher metal. Around 700 AD, the Japanese began folding bloomery-steel and cast-iron in alternating layers to increase the strength of their swords, using clay fluxes to remove slag and impurities. This method of Japanese swordsmithing produced one of the purest steel-alloys of the early Middle Ages.

While the use of iron started to become more widespread around 1200 BC, mainly because of interruptions in the trade routes for tin, the metal was much softer than bronze. However, very small amounts of steel, (an alloy of iron and around 1% carbon), was always a byproduct of the bloomery process. The ability to modify the hardness of steel by heat treatment had been known since 1100 BC, and the rare material was valued for the manufacture of tools and weapons. Because the ancients could not produce temperatures high enough to melt iron fully, the production of steel in decent quantities did not occur until the introduction of blister steel during the Middle Ages. This method introduced carbon by heating wrought iron in charcoal for long periods of time, but the penetration of carbon was not very deep, so the alloy was not homogeneous. In 1740, Benjamin Huntsman began melting blister steel in a crucible to even out the carbon content, creating the first process for the mass production of tool steel. Huntsman›s process was used for manufacturing tool steel until the early 1900s.

With the introduction of the blast furnace to Europe in the Middle Ages, pig iron was able to be produced in much higher volumes than wrought iron. Because pig iron could be melted, people began to develop processes of reducing the carbon in the liquid pig iron to create steel. Puddling was introduced during the 1700s, where molten pig iron was stirred while exposed to the air, to remove the carbon by oxidation. In 1858, Sir Henry Bessemer developed a process of steel-making by blowing hot air through liquid pig iron to reduce the carbon content. The Bessemer process was able to produce the first large scale manufacture of steel. Once the Bessemer process began to gain widespread use, other alloys of steel began to follow. Mangalloy, an alloy of steel and manganese exhibiting extreme hardness and toughness, was one of the first alloy steels, and was created by Robert Hadfield in 1882.

Precipitation-Hardening Alloys

In 1906, precipitation hardening alloys were discovered by Alfred Wilm. Precipitation hardening alloys, such as certain alloys of aluminium, titanium, and copper, are heat-treatable alloys that soften when quenched (cooled quickly), and then harden over time. After quenching a ternary alloy of aluminium, copper, and magnesium, Wilm discovered that the alloy increased in hardness when left to age at room temperature. Although an explanation for the phenomenon was not provided until 1919, duralumin was one of the first "age hardening" alloys to be used, and was soon followed by many others. Because they often exhibit a combination of high strength and low weight, these alloys became widely used in many forms of industry, including the construction of modern aircraft.

REFERENES

1. Callister, W. D. "Materials Science and Engineering: An Introduction" 2007, 7th edition, John Wiley and Sons, Inc. New York, Section 4.3 and Chapter 9.
2. Bauccio, Michael (1003) ASM metals reference book. ASM International. ISBN 0-87170-478-1.
3. Verhoeven, John D. (2007). Steel Metallurgy for the Non-metallurgist. ASM International. p. 56. ISBN 978-1-61503-056-9.
4. Davis, Joseph R. (1993) ASM Specialty Handbook: Aluminum and Aluminum Alloys. ASM International. p. 211. ISBN 978-0-87170-496-2.
5. Mills, Adelbert Phillo (1922) Materials of Construction: Their Manufacture and Properties, John Wiley & sons, inc, originally published by the University of Wisconsin, Madison
6. Hogan, C. (1969). «Density of States of an Insulating Ferromagnetic Alloy». Physical Review. 188 (2): 870. Bibcode:1969PhRv..188..870H. doi:10.1103/PhysRev.188.870.
7. Zhang, X.; Suhl, H. (1985). «Spin-wave-related period doublings and chaos under transverse pumping». Physical Review A. 32 (4): 2530–2533. Bibcode:1985PhRvA..32.2530Z. doi:10.1103/PhysRevA.32.2530. PMID 9896377.
8. Dossett, Jon L. and Boyer, Howard E. (2006) Practical heat treating. ASM International. pp. 1–14. ISBN 1-61503-110-3.
9. Rickard, T. A. (1941). «The Use of Meteoric Iron». Journal of the Royal Anthropological Institute. Royal Anthropological Institute of Great Britain and Ireland. 71 (1/2): 55–66. doi:10.2307/2844401. JSTOR 2844401.
10. Buchwald, pp. 13–22
11. Buchwald, pp. 35–37
12. Buchwald, pp. 39–41
13. Smith, Cyril (1960) History of metallography. MIT Press. pp. 2–4. ISBN 0-262-69120-5.

14. Emperor›s Ghost Army. pbs.org. November 2014
15. Rapp, George (2009) Archaeomineralogy. Springer. p. 180. ISBN 3-540-78593-0
16. Miskimin, Harry A. (1977) The economy of later Renaissance Europe, 1460–1600. Cambridge University Press. p. 31. ISBN 0-521-29208-5.
17. Nicholson, Paul T. and Shaw, Ian (2000) Ancient Egyptian materials and technology. Cambridge University Press. pp. 164–167. ISBN 0-521-45257-0.
18. Kay, Melvyn (2008) Practical Hydraulics. Taylor and Francis. p. 45. ISBN 0-415-35115-4.
19. Hull, Charles (1992) Pewter. Shire Publications. pp. 3–4; ISBN 0-7478-0152-5
20. Brinkley, Frank (1904) Japan and China: Japan, its history, arts, and literature. Oxford University. p. 317
21. Roberts, George Adam; Krauss, George; Kennedy, Richard and Kennedy, Richard L. (1998) Tool steels. ASM International. pp. 2–3. ISBN 0-87170-599-0.
22. Bramfitt, B. L. (2001). Metallographer›s Guide: Practice and Procedures for Irons and Steels. ASM International. pp. 13–. ISBN 978-1-61503-146-7.
23. Jacobs, M. H. Precipitation Hardnening. University of Birmingham. TALAT Lecture 1204. slideshare.net

CHAPTER 8

Metallurgy

Chapter Contents

- **Introduction**
- **Etymology and Pronunciation**
- **History**
- **Alloys**
- **Production**
- **Microstructure**
- **Thermal Spraying**
- **Experimental Archaeometallurgy**
- **Smelting**

In its general, modern sense, metallurgy is the science that studies the chemical and physical properties of metals, including how they perform when used for culturally useful industrial purposes. The term often refers to the procedures used in extracting metals from ore, as well as to the processes related to metals purification and alloy production. It also refers to the craft of making culturally useful objects out of metal, or metalworking. The practice of metalworking has been carried out over thousands of centuries.

INTRODUCTION

Metallurgy is a domain of materials science and engineering that studies the physical and chemical behaviour of metallic elements, their intermetallic compounds, and their mixtures, which are called alloys. Metallurgy is also the technology of metals: the way in which science is applied to the production of metals, and the engineering of metal components for usage in products for consumers and manufacturers. The production of metals involves the processing of ores to extract the metal they contain, and the mixture of metals, sometimes with other elements, to produce alloys. Metallurgy is distinguished from the craft of metalworking, although metalworking relies on metallurgy, as medicine relies on medical science, for technical advancement.

Smelting Gold in Nicaragua in the La Luz Gold Mine in Siuna and Bonanza about 1959. Smelting is a basic step in obtaining usable quantities of most metals.

Pouring Smelted Gold into an ingot at the La Luz Gold Mine in Siuna, Nicaragua about 1959.

Metallurgy is subdivided into *ferrous metallurgy* (sometimes also known as black metallurgy) and *non-ferrous metallurgy* or *colored metallurgy*. Ferrous metallurgy involves processes and alloys based on iron while non-ferrous metallurgy involves processes and alloys based on other metals. The production of ferrous metals accounts for 95 percent of world metal production.

ETYMOLOGY AND PRONUNCIATION

The roots of *metallurgy* derive from Ancient Greek: μεταλλουργός, *metallourgós*, "worker in metal", from μέταλλον, *métallon*, "metal" + ργον, *érgon*, work.

The word was originally an alchemist›s term for the extraction of metals from minerals, the ending *-urgy* signifying a process, especially manufacturing: it was discussed in this sense in the 1797 Encyclopaedia Britannica. In the late 19th century it was extended to the more general scientific study of metals, alloys, and related processes.

In English, the /metælədi/ pronunciation is the more common one in the UK and Commonwealth. The /metələrdi/ pronunciation is the more common one in the USA, and is the first-listed variant in various American dictionaries (e.g., *Merriam-Webster Collegiate*, *American Heritage*).

HISTORY

Gold headband from Thebes 750–700 BC.

The earliest recorded metal employed by humans appears to be gold, which can be found free or "native". Small amounts of natural gold have been found in Spanish caves used during the late Paleolithic period, c. 40,000 BC. Silver, copper, tin and meteoric iron can also be found in native form, allowing a limited amount of metalworking in early cultures. Egyptian weapons made from meteoric iron in about 3000 BC were highly prized as "daggers from heaven".

Certain metals, notably tin, lead and (at a higher temperature) copper, can be recovered from their ores by simply heating the rocks in a fire or blast furnace, a process known as smelting. The first evidence of this extractive metallurgy dates from the 5th and 6th millennia BC and was found in the archaeological sites of Majdanpek, Yarmovac and Plocnik, all three in Serbia. To date, the earliest evidence of copper smelting is found at the Belovode site, including a copper axe from 5500 BC belonging to the Vinča culture. Other signs of early metals are found from the third millennium BC in places like Palmela (Portugal), Los Millares (Spain), and Stonehenge (United Kingdom). However, the ultimate beginnings cannot be clearly ascertained and new discoveries are both continuous and ongoing.

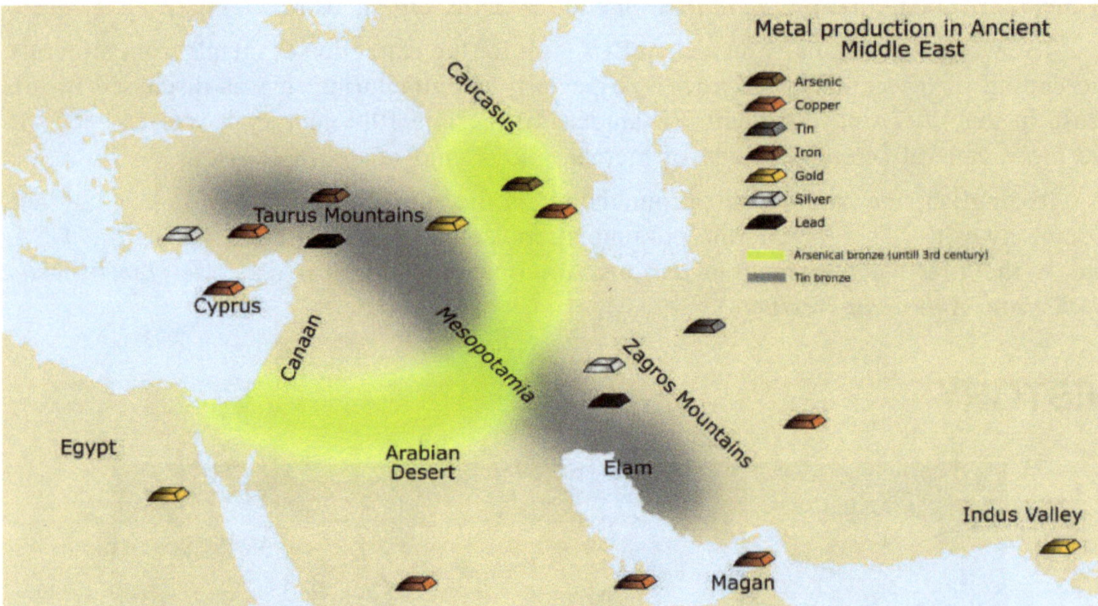

Mining areas of the ancient Middle East. Boxes colors: arsenic is in brown, copper in red, tin in grey, iron in reddish brown, gold in yellow, silver in white and lead in black. Yellow area stands for arsenic bronze, while grey area stands for tin bronze.

These first metals were single ones or as found. About 3500 BC, it was discovered that by combining copper and tin, a superior metal could be made, an alloy called bronze, representing a major technological shift known as the Bronze Age.

The extraction of iron from its ore into a workable metal is much more difficult than for copper or tin. The process appears to have been invented by the Hittites in about 1200 BC, beginning the Iron Age. The secret of extracting and working iron was a key factor in the success of the Philistines.

Historical developments in ferrous metallurgy can be found in a wide variety of past cultures and civilizations. This includes the ancient and medieval kingdoms and empires of the Middle East and Near East, ancient Iran, ancient Egypt, ancient Nubia, and Anatolia (Turkey), Ancient Nok, Carthage, the Greeks and Romans of ancient Europe, medieval Europe, ancient and medieval China, ancient and medieval India, ancient and medieval Japan, amongst others. Many applications, practices, and devices associated or involved in metallurgy were established in ancient China, such as the innovation of the blast furnace, cast iron, hydraulic-powered trip hammers, and double acting piston bellows.

A 16th century book by Georg Agricola called *De re metallica* describes the highly developed and complex processes of mining metal ores, metal extraction and metallurgy of the time. Agricola has been described as the "father of metallurgy".

Extraction

Furnace bellows operated by waterwheels, Yuan Dynasty, China.

Aluminium plant in Žiar nad Hronom (Central Slovakia).

Extractive metallurgy is the practice of removing valuable metals from an ore and refining the extracted raw metals into a purer form. In order to convert a metal oxide or sulfide to a purer metal, the ore must be reduced physically, chemically, or electrolytically.

Extractive metallurgists are interested in three primary streams: feed, concentrate (valuable metal oxide/sulfide), and tailings (waste). After mining, large pieces of the ore feed are broken through crushing and/or grinding in order to obtain particles small enough where each particle is either mostly valuable or mostly waste. Concentrating the particles of value in a form supporting separation enables the desired metal to be removed from waste products.

Mining may not be necessary if the ore body and physical environment are conducive to leaching. Leaching dissolves minerals in an ore body and results in an enriched solution. The solution is collected and processed to extract valuable metals.

Ore bodies often contain more than one valuable metal. Tailings of a previous process may be used as a feed in another process to extract a secondary product from the original ore. Additionally, a concentrate may contain more than one valuable metal. That concentrate would then be processed to separate the valuable metals into individual constituents.

Alloys

Casting bronze.

Common engineering metals include aluminium, chromium, copper, iron, magnesium, nickel, titanium and zinc. These are most often used as alloys. Much effort has been placed on understanding the iron-carbon alloy system, which includes steels and cast irons. Plain carbon steels (those that contain essentially only carbon as an alloying element) are used in low-cost, high-strength applications where weight and corrosion are not a problem. Cast irons, including ductile iron, are also part of the iron-carbon system.

Stainless steel or galvanized steel are used where resistance to corrosion is important. Aluminium alloys and magnesium alloys are used for applications where strength and lightness are required.

Copper-nickel alloys (such as Monel) are used in highly corrosive environments and for non-magnetic applications. Nickel-based superalloys like Inconel are used in high-temperature applications such as gas turbines, turbochargers, pressure vessels, and heat exchangers. For extremely high temperatures, single crystal alloys are used to minimize creep.

PRODUCTION

In production engineering, metallurgy is concerned with the production of metallic components for use in consumer or engineering products. This involves the production of alloys, the shaping, the heat treatment and the surface treatment of the product. The task of the metallurgist is to achieve balance between material properties such as cost, weight, strength, toughness, hardness, corrosion, fatigue resistance, and performance in temperature extremes. To achieve this goal, the operating environment must be carefully considered. In a saltwater environment, ferrous metals and some aluminium alloys corrode quickly. Metals exposed to cold or cryogenic conditions may endure a ductile to brittle transition and lose their toughness, becoming more brittle and prone to cracking. Metals under continual cyclic loading can suffer from metal fatigue. Metals under constant stress at elevated temperatures can creep.

Metalworking Processes

Metals are shaped by processes such as:

- casting – molten metal is poured into a shaped mold.
- forging – a red-hot billet is hammered into shape.
- rolling – a billet is passed through successively narrower rollers to create a sheet.
- laser cladding – metallic powder is blown through a movable laser beam (e.g. mounted on a NC 5-axis machine). The resulting melted metal reaches a substrate to form a melt pool. By moving the laser head, it is possible to stack the tracks and build up a three-dimensional piece.
- extrusion – a hot and malleable metal is forced under pressure through a die, which shapes it before it cools.
- sintering – a powdered metal is heated in a non-oxidizing environment after being compressed into a die.
- machining – lathes, milling machines, and drills cut the cold metal to shape.
- fabrication – sheets of metal are cut with guillotines or gas cutters and bent and welded into structural shape.
- 3D printing – Sintering or melting powder metal in a very small point on a moving ‘print head’ moving in 3D space to make any object to shape.

Cold-working processes, in which the product’s shape is altered by rolling, fabrication or other processes while the product is cold, can increase the strength of the product by a process called work hardening. Work hardening creates microscopic defects in the metal, which resist further changes of shape.

Various forms of casting exist in industry and academia. These include sand casting, investment casting (also called the "lost wax process), die casting, and continuous casting.

Heat Treatment

Metals can be heat-treated to alter the properties of strength, ductility, toughness, hardness and/or resistance to corrosion. Common heat treatment processes include annealing, precipitation strengthening, quenching, and tempering. The **annealing** process softens the metal by heating it and then allowing it to cool very slowly, which gets rid of stresses in the metal and makes the grain structure large and soft-edged so that when the metal is hit or stressed it dents or perhaps bends, rather than breaking; it is also easier to sand, grind, or cut annealed metal. **Quenching** is the process of cooling a high-carbon steel very quickly after heating, thus "freezing" the steel's molecules in the very hard martensite form, which makes the metal harder. There is a balance between hardness and toughness in any steel; the harder the steel, the less tough or impact-resistant it is, and the more impact-resistant it is, the less hard it is. **Tempering** relieves stresses in the metal that were caused by the hardening process; tempering makes the metal less hard while making it better able to sustain impacts without breaking.

Often, mechanical and thermal treatments are combined in what are known as thermo-mechanical treatments for better properties and more efficient processing of materials. These processes are common to high-alloy special steels, super alloys and titanium alloys.

Plating

Electroplating is a chemical surface-treatment technique. It involves bonding a thin layer of another metal such as gold, silver, chromium or zinc to the surface of the product. It is used to reduce corrosion as well as to improve the product's aesthetic appearance.

Thermal Spraying

Thermal spraying techniques are another popular finishing option, and often have better high temperature properties than electroplated coatings.

MICROSTRUCTURE

Metallurgists study the microscopic and macroscopic properties using metallography, a technique invented by Henry Clifton Sorby. In metallography, an alloy of interest is ground flat and polished to a mirror finish. The sample can then be etched to reveal the microstructure and macrostructure of the metal. The sample is then examined in an optical or electron microscope, and the image contrast provides details on the composition, mechanical properties, and processing history.

Crystallography, often using diffraction of x-rays or electrons, is another valuable tool available to the modern metallurgist. Crystallography allows identification of unknown materials and reveals the crystal structure of the sample. Quantitative crystallography can be used to calculate the amount of phases present as well as the degree of strain to which a sample has been subjected.

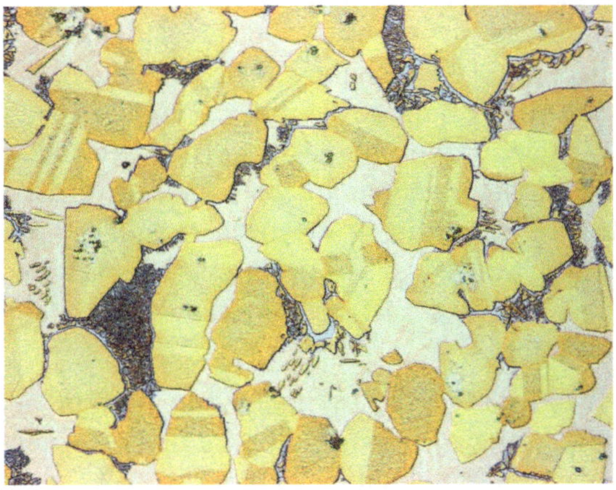

Metallography allows the metallurgist to study the microstructure of metals.

Conferences

EMC, the European Metallurgical Conference has developed to the most important networking business event dedicated to the non-ferrous metals industry in Europe. From the start of the conference sequence in 2001 at Friedrichshafen it was host of the most relevant metallurgists from all countries of the world. The European Metallurgical Conference is organized by GDMB Society of Metallurgists and Miners.

THERMAL SPRAYING

Plasma spraying setup – a variant of thermal spraying.

Thermal spraying techniques are coating processes in which melted (or heated) materials are sprayed onto a surface. The "feedstock" (coating precursor) is heated by electrical (plasma or arc) or chemical means (combustion flame).

Thermal spraying can provide thick coatings (approx. thickness range is 20 micrometers to several mm, depending on the process and feedstock), over a large area at high deposition rate as compared to other coating processes such as electroplating, physical and chemical vapor deposition. Coating materials available for thermal spraying include metals, alloys, ceramics, plastics and composites. They are fed in powder or wire form, heated to a molten or semimolten state and accelerated towards substrates in the form of micrometer-size particles. Combustion or electrical arc discharge is usually used as the source of energy for thermal spraying. Resulting coatings are made by the accumulation of numerous sprayed particles. The surface may not heat up significantly, allowing the coating of flammable substances.

Coating quality is usually assessed by measuring its porosity, oxide content, macro and micro-hardness, bond strength and surface roughness. Generally, the coating quality increases with increasing particle velocities.

Several variations of thermal spraying are distinguished:

- Plasma spraying
- Detonation spraying
- Wire arc spraying
- Flame spraying
- High velocity oxy-fuel coating spraying (HVOF)
- High velocity air fuel (HVAF)
- Warm spraying
- Cold spraying

In classical (developed between 1910 and 1920) but still widely used processes such as flame spraying and wire arc spraying, the particle velocities are generally low (< 150 m/s), and raw materials must be molten to be deposited. Plasma spraying, developed in the 1970s, uses a high-temperature plasma jet generated by arc discharge with typical temperatures >15000 K, which makes it possible to spray refractory materials such as oxides, molybdenum, etc.

System Overview

A typical thermal spray system consists of the following:

- Spray torch (or spray gun) – the core device performing the melting and acceleration of the particles to be deposited
- Feeder – for supplying the powder, wire or liquid to the torch through tubes.
- Media supply – gases or liquids for the generation of the flame or plasma jet, gases for carrying the powder, etc.
- Robot – for manipulating the torch or the substrates to be coated

- Power supply – often standalone for the torch
- Control console(s) – either integrated or individual for all of the above

Detonation Thermal Spraying Process

The detonation gun consists of a long water-cooled barrel with inlet valves for gases and powder. Oxygen and fuel (acetylene most common) is fed into the barrel along with a charge of powder. A spark is used to ignite the gas mixture, and the resulting detonation heats and accelerates the powder to supersonic velocity through the barrel. A pulse of nitrogen is used to purge the barrel after each detonation. This process is repeated many times a second. The high kinetic energy of the hot powder particles on impact with the substrate results in a buildup of a very dense and strong coating.

Plasma Spraying

Wire flame spraying.

In plasma spraying process, the material to be deposited (feedstock) — typically as a powder, sometimes as a liquid, suspension or wire — is introduced into the plasma jet, emanating from a plasma torch. In the jet, where the temperature is on the order of 10,000 K, the material is melted and propelled towards a substrate. There, the molten droplets flatten, rapidly solidify and form a deposit. Commonly, the deposits remain adherent to the substrate as coatings; free-standing parts can also be produced by removing the substrate. There are a large number of technological parameters that influence the interaction of the particles with the plasma jet and the substrate and therefore the deposit properties. These parameters include feedstock type, plasma gas composition and flow rate, energy input, torch offset distance, substrate cooling, etc.

Deposit Properties

The deposits consist of a multitude of pancake-like 'splats' called lamellae, formed by flattening of the liquid droplets. As the feedstock powders typically have sizes from micrometers to above 100 micrometers, the lamellae have thickness in the micrometer range and lateral dimension from several to hundreds of micrometers. Between these lamellae, there are small voids, such as pores, cracks and regions of incomplete bonding. As a result of this unique structure, the deposits can have properties significantly different from bulk materials. These are generally mechanical properties, such as lower strength and modulus, higher strain tolerance, and lower thermal and electrical conductivity. Also, due to the rapid solidification, metastable phases can be present in the deposits.

Applications

This technique is mostly used to produce coatings on structural materials. Such coatings provide protection against high temperatures (for example thermal barrier coatings for exhaust heat management), corrosion, erosion, wear; they can also change the appearance, electrical or tribological properties of the surface, replace worn material, etc. When sprayed on substrates of various shapes and removed, free-standing parts in the form of plates, tubes, shells, etc. can be produced. It can also be used for powder processing (spheroidization, homogenization, modification of chemistry, etc.). In this case, the substrate for deposition is absent and the particles solidify during flight or in a controlled environment (e.g., water). This technique with variation may also be used to create porous structures, suitable for bone ingrowth, as a coating for medical implants. A polymer dispersion aerosol can be injected into the plasma discharge in order to create a grafting of this polymer on to a substrate surface. This application is mainly used to modify the surface chemistry of polymers.

Variations

Plasma spraying systems can be categorized by several criteria.

Plasma jet generation

- direct current (DC plasma), where the energy is transferred to the plasma jet by a direct current, high-power electric arc
- induction plasma or RF plasma, where the energy is transferred by induction from a coil around the plasma jet, through which an alternating, radio-frequency current passes

Plasma-forming medium

- gas-stabilized plasma (GSP), where the plasma forms from a gas; typically argon, hydrogen, helium or their mixtures
- water-stabilized plasma (WSP), where plasma forms from water (through evaporation, dissociation and ionization) or other suitable liquid
- hybrid plasma – with combined gas and liquid stabilization, typically argon and water

Spraying environment

- air plasma spraying (APS), performed in ambient air
- controlled atmosphere plasma spraying (CAPS), usually performed in a closed chamber, either filled with inert gas or evacuated
- variations of CAPS: high-pressure plasma spraying (HPPS), low-pressure plasma spraying (LPPS), the extreme case of which is vacuum plasma spraying (VPS, see below)
- underwater plasma spraying

Another variation consists of having a liquid feedstock instead of a solid powder for melting, this technique is known as Solution precursor plasma spray

Vacuum Plasma Spraying

Vacuum plasma spraying.

Vacuum plasma spraying (VPS) is a technology for etching and surface modification to create porous layers with high reproducibility and for cleaning and surface engineering of plastics, rubbers and natural fibers as well as for replacing CFCs for cleaning metal components. This surface engineering can improve properties such as frictional behavior, heat resistance, surface electrical conductivity, lubricity, cohesive strength of films, or dielectric constant, or it can make materials hydrophilic or hydrophobic.

The process typically operates at 39–120 °C to avoid thermal damage. It can induce non-thermally activated surface reactions, causing surface changes which cannot occur with molecular chemistries at atmospheric pressure. Plasma processing is done in a controlled environment inside a sealed chamber at a medium vacuum, around 13–65 Pa. The gas or mixture of gases is energized by an electrical field from DC to microwave frequencies, typically 1–500 W at 50 V. The treated components are usually electrically isolated. The volatile plasma by-products are evacuated from the chamber by the vacuum pump, and if necessary can be neutralized in an exhaust scrubber.

In contrast to molecular chemistry, plasmas employ:

- Molecular, atomic, metastable and free radical species for chemical effects.
- Positive ions and electrons for kinetic effects.

Plasma also generates electromagnetic radiation in the form of vacuum UV photons to penetrate bulk polymers to a depth of about 10 μm. This can cause chain scissions and cross-linking.

Plasmas affect materials at an atomic level. Techniques like X-ray photoelectron spectroscopy and scanning electron microscopy are used for surface analysis to identify the processes required and to judge their effects. As a simple indication of surface energy, and hence adhesion or wettability, often a water droplet contact angle test is used. The lower the contact angle, the higher the surface energy and more hydrophilic the material is.

Changing Effects with Plasma

At higher energies ionization tends to occur more than chemical dissociations. In a typical reactive gas, 1 in 100 molecules form free radicals whereas only 1 in 10^6 ionizes. The predominant effect here is the forming of free radicals. Ionic effects can predominate with selection of process parameters and if necessary the use of noble gases.

Wire arc spray

Wire arc spray is a form of thermal spraying where two consumable metal wires are fed independently into the spray gun. These wires are then charged and an arc is generated between them. The heat from this arc melts the incoming wire, which is then entrained in an air jet from the gun. This entrained molten feedstock is then deposited onto a substrate. This process is commonly used for metallic, heavy coatings.

Plasma Transferred Wire Arc

Plasma transferred wire arc is another form of wire arc spray which deposits a coating on the internal surface of a cylinder, or on the external surface of a part of any geometry. It is predominantly known for its use in coating the cylinder bores of an engine, enabling the use of Aluminum engine blocks without the need for heavy cast iron sleeves. A single conductive wire is used as "feedstock" for the system. A supersonic plasma jet melts the wire, atomizes it and propels it onto the substrate. The plasma jet is formed by a transferred arc between a non-consumable cathode and the type of a wire. After atomization, forced air transports the stream of molten droplets onto the bore wall. The particles flatten when they impinge on the surface of the substrate, due to the high kinetic energy. The particles rapidly solidify upon contact. The stacked particles make up a high wear resistant coating. The PTWA thermal spray process utilizes a single wire as the feedstock material. All conductive wires up to and including 0.0625" (1.6mm) can be used as feedstock material, including "cored" wires. PTWA can be used to apply a coating to the wear surface of engine or transmission components to replace a bushing or bearing. For example, using PTWA to coat the bearing surface of a connecting rod offers a number of benefits including reductions in weight, cost, friction potential, and stress in the connecting rod.

High velocity oxygen fuel spraying (HVOF)

During the 1980s, a class of thermal spray processes called high velocity oxy-fuel spraying was developed. A mixture of gaseous or liquid fuel and oxygen is fed into a combustion chamber, where they are ignited and combusted continuously. The resultant hot gas at a pressure close to 1 MPa emanates through a converging–diverging nozzle and travels through a straight section. The fuels can be gases (hydrogen, methane, propane, propylene, acetylene, natural gas, etc.) or liquids (kerosene, etc.). The jet velocity at the exit of the barrel (>1000 m/s) exceeds the speed of sound. A powder feed stock is injected into the gas stream, which accelerates the powder up to 800 m/s. The stream of hot gas and powder is directed towards the surface to be coated. The powder partially melts in the stream, and deposits upon the substrate. The resulting coating has low porosity and high bond strength.

HVOF coatings may be as thick as 12 mm (1/2). It is typically used to deposit wear and corrosion resistant coatings on materials, such as ceramic and metallic layers. Common powders include WC-Co, chromium carbide, MCrAlY, and alumina. The process has been most successful for depositing cermet materials (WC–Co, etc.) and other corrosion-resistant alloys (stainless steels, nickel-based alloys, aluminium, hydroxyapatite for medical implants, etc.).

Cold Spraying

In the 1990s, cold spraying (often called gas dynamic cold spray) was introduced. The method was originally developed in Russia, with the accidental observation of the rapid formation of coatings. This occurred while experimenting with particle erosion of a target exposed to a high velocity flow loaded with fine powder in a wind tunnel. In cold spraying, particles are accelerated to very high speeds by the carrier gas forced through a converging–diverging de Laval type nozzle. Upon impact, solid particles with sufficient kinetic energy deform plastically and bond mechanically to the substrate to form a coating. The critical velocity needed to form bonding depends on the material's properties, powder size and temperature. Metals, polymers, ceramics, composite materials and nanocrystalline powders can be deposited using cold spraying. Soft metals such as

Cu and Al are best suited for cold spraying, but coating of other materials (W, Ta, Ti, MCrAlY, WC–Co, etc.) by cold spraying has been reported.

The deposition efficiency is typically low for alloy powders, and the window of process parameters and suitable powder sizes is narrow. To accelerate powders to higher velocity, finer powders (<20 micrometers) are used. It is possible to accelerate powder particles to much higher velocity using a processing gas having high speed of sound (helium instead of nitrogen). However, helium is costly and its flow rate, and thus consumption, is higher. To improve acceleration capability, nitrogen gas is heated up to about 900 C. As a result, deposition efficiency and tensile strength of deposits increase.

Warm Spraying

Warm spraying is a novel modification of high velocity oxy-fuel spraying, in which the temperature of combustion gas is lowered by mixing nitrogen with the combustion gas, thus bringing the process closer to the cold spraying. The resulting gas contains much water vapor, unreacted hydrocarbons and oxygen, and thus is dirtier than the cold spraying. However, the coating efficiency is higher. On the other hand, lower temperatures of warm spraying reduce melting and chemical reactions of the feed powder, as compared to HVOF. These advantages are especially important for such coating materials as Ti, plastics, and metallic glasses, which rapidly oxidize or deteriorate at high temperatures.

Applications

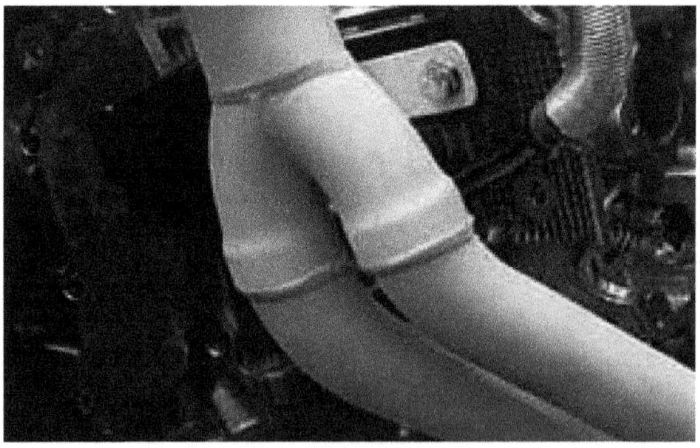

Plasma sprayed ceramic coating applied onto a part of an automotive exhaust system.

- Crankshaft reconditioning or conditioning
- Corrosion protection
- Fouling protection
- Altering thermal conductivity or electrical conductivity
- Wear control: either hardfacing (wear-resistant) or abradable coating
- Repairing damaged surfaces

- Temperature/oxidation protection (thermal barrier coatings)
- Medical implants
- Production of functionally graded materials (for any of the above applications)

Limitations

Thermal spraying is a line of sight process and the bond mechanism is primarily mechanical. Thermal spray application is not compatible with the substrate if the area to which it is applied is complex or blocked by other bodies.

Safety

Thermal spraying need not be a dangerous process, if the equipment is treated with care, and correct spraying practices are followed. As with any industrial process, there are a number of hazards, of which the operator should be aware, and against which specific precautions should be taken. Ideally, equipment should be operated automatically, in enclosures specially designed to extract fumes, reduce noise levels, and prevent direct viewing of the spraying head. Such techniques will also produce coatings that are more consistent. There are occasions when the type of components being treated, or their low production levels, requires manual equipment operation. Under these conditions, a number of hazards, peculiar to thermal spraying, are experienced, in addition to those commonly encountered in production or processing industries.

Noise

Metal spraying equipment uses compressed gases, which create noise. Sound levels vary with the type of spraying equipment, the material being sprayed, and the operating parameters. Typical sound pressure levels are measured at 1 meter behind the arc.

UV light

Combustion spraying equipment produces an intense flame, which may have a peak temperature more than 3,100 °C, and is very bright. Electric arc spraying produces ultra-violet light, which may damage delicate body tissues. Spray booths, and enclosures, should be fitted with ultra-violet absorbent dark glass. Where this is not possible, operators, and others in the vicinity should wear protective goggles containing BS grade 6 green glass. Opaque screens should be placed around spraying areas. The nozzle of an arc pistol should never be viewed directly, unless it is certain that no power is available to the equipment.

Dust and Fumes

The atomization of molten materials produces a large amount of dust and fumes made up of very fine particles (ca. 80–95% of the particles by number <100 nm). Proper extraction facilities are vital, not only for personal safety, but to minimize entrapment of re-frozen particles in the sprayed coatings. The use of respirators, fitted with suitable filters, is strongly recommended, where equipment cannot be isolated. Certain materials offer specific known hazards:

- Finely divided metal particles are potentially pyrophoric and harmful when accumulated in the body.
- Certain materials e.g. aluminum, zinc and other base metals may react with water to evolve hydrogen. This is potentially explosive and special precautions are necessary in fume extraction equipment.
- Fumes of certain materials, notably zinc and copper alloys, have a disagreeable odour and may cause a fever-type reaction in certain individuals (known as metal fume fever). This may occur some time after spraying and usually subsides rapidly. If it does not, medical advice must be sought.
- Fumes of reactive compounds can dissociate and create harmful gasses. Respirators should be worn in these areas and gas meters should be used to monitor the air before respirators are removed.

Heat

Combustion spraying guns use oxygen and fuel gases. The fuel gases are potentially explosive. In particular, acetylene may only be used under approved conditions. Oxygen, while not explosive, will sustain combustion, and many materials will spontaneously ignite, if excessive oxygen levels are present. Care must be taken to avoid leakage, and to isolate oxygen and fuel gas supplies, when not in use.

Shock Hazards

Electric arc guns operate at low voltages (below 45 V dc), but at relatively high currents. They may be safely hand-held. The power supply units are connected to 440 V AC sources, and must be treated with caution.

EXPERIMENTAL ARCHAEOMETALLURGY

Experimental archaeometallurgy is a subset of experimental archaeology that specifically involves past metallurgical processes most commonly involving the replication of copper and iron objects as well as testing the methodology behind the production of ancient metals and metal objects. Metals and elements used primarily as alloying materials, such as tin, lead, and arsenic, are also a part of experimental research.

Experimental Archaeometallurgy as a Sub-Discipline

The theory behind experimental archaeology comes from the new archaeology technique of the 1950s to use modern day examples in the form of experiments and ethnologies as analogues to past processes. Experimental archaeometallurgy is considered a part of general experimental archaeology and is rarely separated in the literature and as such, many of the principles stay the same while there is a greater focus on a single subject.

Archaeometallurgy works as a good field for experimental reproduction because of the evidence that is provided from excavation is a good starting point for reconstruction. Metallurgical remains provide a durable product that has relatively durable evidence of production methods such as slag and refractory ceramic remains. Experimentation comes in a varied amount of forms including object replication, system replication, behavioral replication, and process replication.

Research Problems

Archaeometallurgical experimentation typically takes place in controlled laboratories or tries to remain as authentic as possible by being conducted using only the materials and facilities that were available to the subjects whose technology is trying to be reconstructed. Regardless of location though, the experimentation is always conducted under a different mindset outside the context of what was originally intended. A constant problem in any type of experimental archaeology is the cultural distance between the archaeologist and the individual who originally was involved with the metallurgy. This difference in mindset may lead to misunderstandings in the processes behind the metallurgy. Second to this, not all experiments are successful and it is hard to determine if this is the fault of the techniques used or the individual conducting the experiment.

Connection with Ethnoarchaeology

Ethnoarchaeology has been widely used in conjunction with experimental archaeology using the techniques of modern peoples as analogues to the processes of the past. The attempted use of ethnology in archaeology tries to counteract the cultural distance of the researcher from process by changing the context of experimentation. Africa has played a large role in reconstructing copper smelting and bloomery iron furnaces as there are still several places that practice a workshop production of iron. Killick has been one such archaeologist to utilize surviving iron production in Africa to gain further insight into how other furnaces from around the world may have been constructed.

Mining

Mining is among the first steps of producing metal and as such is one of the foci of experimental archaeometallurgy. However, experimental research on mining is mostly limited to firesetting and the reproduction and use of mining tools.

Firesetting

Firesetting is the process of exposing a rock face to high temperatures to induce cracking, spalling, and an overall increase to the brittleness of the rock in order to make it more susceptible to mining processes. Understanding the process of firesetting has been a crucial element to the development of an archaeological history of mining and as such has been the subject of several experiments to reproduce the technique. Typically firesetting experiments are conducted by setting a fire next to a predetermined rock face while taking measurements on the amount of and type of fuel used, temperatures of the fire and rock face, the amount of spalling before and after excavation, as well as the amount of time required for the different procedures. This examination allows for several possible inferences to be made about the mining process including the total amount of fuel a mining site may have needed to complete and its effects on the surrounding environment as well as how mining labor could have been organized. One outcome of firesetting experimentation is the realization that the quenching, or dousing the rock face with water after heating, is not necessary to making the rock face easier to excavate. The quenching process had been a standard step in most experiments with firesetting, but now more research is necessary to answer the new question of why quenching was used if it was not effective.

Mining Tools

Experimental reconstruction of tools used in prehistoric mining is often written about in conjunction with the tools use after the process of firesetting. The experimental mining tool assemblage are primarily made up of hammerstones and antler picks that are reconstructed using willow and hazel sticks, rawhide, and hemp string to implement various hafting techniques and methods of utilization.

Smelting

Smelting or the reduction of an ore to its metallic state is the primary source of experimentation in archaeometallurgy. In its simplest form smelting can be accomplished by placing an ore sample between two pieces of combusting charcoal in an oxygen reducing atmosphere with a compressed air source to feed the combustion and result in temperatures high enough to smelt metal. But to reach this final metallic state several things need to be done first including the processing of the ore to remove waste or gangue material, the possible roasting of the ore, the smelting of the ore, and then there is the possibility of refining the metal through a series of remelts. Then, through chemical or microscopic analysis, the products of the smelt are analyzed and compared with the findings of archaeological excavation in order to examine the likelihood of various manufacturing processes.

Copper and its alloys

The first recorded experimental work in copper was conducted by Cushing in 1894 in order to demonstrate that copper plate found in Hopewell mounds in Ohio could have been produced by the Hopewell people and not come from European trade. In his experiments, Cushing used antler and stone tools to cut out sheets of copper and puncture round holes in them through a method of pressure and grinding. In addition to the copper sheet experiments, Cushing also conducted smelting experiments reconstructed from finding at Salado Valley, Arizona pueblo excavations to reproduce copper artifacts.

Iron

The experimental archaeometallurgy of iron is more recent then that of copper in that for the most part was not widely studied until the mid-20th century. This can be attributed to the modern smelting of wrought iron still being produced as an industry up until 1900, when the last of the large-scale production shut down, along with the belief among researchers that many of the same techniques had been passed down since the inception of bloomery iron. A static technique simply was not the case as the technology used to make Roman Era iron showed the use of a technology that had long since disappeared. That being said research in iron has progressed beyond that of copper due to the greater amount of historic text and surviving remains of iron production.

Several experiments have taken place to reproduce bloomery iron. Clough presents an average example of experimental ironwork with the reproduction possible Roman bowl furnaces. Clough found the reproduction bowl furnaces to be inefficient by producing small amounts of poor quality iron, which, when compared to excavated findings of much larger better quality blooms of iron led to the conclusion that bowl furnaces were not used by the Romans.

Precious Metals

Experimental work on precious metals is limited by the cost of experimentation and by the well-understood technical processes involved. Gold and silver are produced in a similar manner to copper with the additional process of cupellation. Platinum is mostly an issue in South America and is typically left out from experimental archaeometallurgy because of its traditional use as a powdered metal as an additive to produce alloys.

Areas of Research

Several sets of data can be collected during the experimental process of smelting including fuel consumption rates, the effects of variation in furnace airflow, temperatures, production time, and chemical composition.

Replication of Techniques

The replication of technique in copper production includes a vast number of possibilities in trying to recreate what has been found through archaeological excavation. Tylecote and Boydell have experimented on possible explanations for the levels of iron found in certain copper objects and the possibility of removing excess iron through the re-melting of the copper. Crew has also done experimental work on iron to show possible loss in iron mass due to the processes involved with working the metal from bloom to billet which concluded with a loss of 75% in slag, impurities, and iron metal.

Alloys

Alloys that are not bronze and brass have had a limited representation in the literature for archaeometallurgy. This is mostly due to lack of interest or evidence in the archaeological record. Arsenical copper is one such limited research topic with some experimental work done by Pollard, Thomas, and Williams. Through several experimental smeltings of copper ores including arsenic, Pollard, Thomas, and Williams found that arsenic in copper is retained in higher levels when a lower smelting temperature is used, implying that arsenical copper may have been the result of early smelting technologies where temperatures where unable to pass a certain point.

Lead experimentation has been limited mostly because of its ease in production. Ore containing lead can be easily smelted, re-melted, and worked and as such there is not much difficultly in understanding how past societies may have produced lead. When lead experiments are conducted, they are done much in the same fashion as copper smelting experiments taking notes on quantitative elements such as completion time, airflow rates, fuel usage, and the resulting amount and composition of metal from the smelt. In addition, lead is a toxic element and special care has to take place in order to experiment with it, which makes limitations on the experiments.

Other Aspects

Artifact replication plays an important role in comparing artifact use. Often objects are made not just to prove a manufacturing process or to sit in a display case, but to show that a given object will show signs of wear that are similar to those present in the archaeological record. Roberts and Ottaway conducted such experimental reconstructions by casting bronze axes using them in a preconceived manner and then comparing the results against known archaeological remains. Results gathered from such experimentation have found that objects have comparable wear patterns and there are European socketed axes that were deposited used as well as unused. Tool mark identification can also go the opposite ways using experimental reconstruction to show the difference between various material media and the wear patterns they leave. Greenfieldgives one such experiment where steel, bronze, and stone tool marks on bone are all researched and examples are given for how they might be seen in the archaeological record.

Energy consumption and efficiency is another topic of interest in archaeometallurgy. Tree felling and land clearing experiments involving comparison of stone, bronze, and steel axes are popular with a number of archaeologists In these types of experiments, factors such as time spent and oxygen intake of the researchers are taken into account to try to find similarities in past life ways use of energy.

SMELTING

Smelting is a form of extractive metallurgy; its main use is to produce a base metal from its ore. This includes production of silver, iron, copper and other base metals from their ores. Smelting makes use of heat and a chemical reducing agent to decompose the ore, driving off other elements as gases or slag and leaving just the metal base behind. The reducing agent is commonly a source of carbon such as coke, or in earlier times charcoal.

The carbon (or carbon monoxide derived from it) removes oxygen from the ore, leaving behind the elemental metal. The carbon is thus oxidized in two stages, producing first carbon monoxide and then carbon dioxide. As most ores are impure, it is often necessary to use flux, such as limestone, to remove the accompanying rock gangue as slag.

Plants for the electrolytic reduction of aluminium are also generally referred to as aluminium smelters.

Electric phosphate smelting furnace in a TVA chemical plant (1942)

Process

Smelting involves more than just melting the metal out of its ore. Most ores are a chemical compound of the metal with other elements, such as oxygen (as an oxide), sulfur (as a sulfide) or carbon and oxygen together (as a carbonate). To produce the metal, these compounds have to undergo a chemical reaction. Smelting therefore consists of using suitable reducing substances that will combine with those oxidizing elements to free the metal.

Roasting

In the case of carbonates and sulfides, a process called "roasting" drives out the unwanted carbon or sulfur, leaving an oxide, which can be directly reduced. Roasting is usually carried out in an oxidizing environment. A few practical examples:

- Malachite, a common ore of copper, is primarily copper carbonate hydroxide $Cu_2(CO_3)(OH)_2$. This mineral undergoes thermal decomposition to $2CuO$, CO_2, and H_2O in several stages between 250 °C and 350 °C. The carbon dioxide and water are expelled into the atmosphere, leaving copper(II) oxide which can be directly reduced to copper as described in the following section titled Reduction.
- Galena, the most common mineral of lead, is primarily lead sulfide (PbS). The sulfide is oxidized to a sulfite ($PbSO_3$) which thermally decomposes into lead oxide and sulfur dioxide gas. (PbO and SO_2) The sulfur dioxide is expelled (like the carbon dioxide in the previous example), and the lead oxide is reduced as below.

Reduction

Reduction is the final, high-temperature step in smelting. It is here that the oxide becomes the elemental metal. A reducing environment (often provided by carbon monoxide, made by incomplete combustion, produced in an air-starved furnace) pulls the final oxygen atoms from the raw metal. The required temperature varies over a very large range, both in absolute terms and in terms of the melting point of the base metal. A few examples:

- iron oxide becomes metallic iron at roughly 1250 °C (2282 °F or 1523.15 K), almost 300 degrees *below* iron's melting point of 1538 °C (2800.4 °F or 1811.15 K)
- mercuric oxide becomes vaporous mercury near 550 °C (1022 °F or 823.15 K), almost 600 degrees *above* mercury's melting point of -38 °C (-36.4 °F or 235.15 K)

Flux and slag can provide a secondary service after the reduction step is complete: They provide a molten cover on the purified metal, preventing it from coming into contact with oxygen while it is still hot enough to oxidize readily.

Fluxes

Fluxes are used in smelting for several purposes, chief among them catalyzing the desired reactions and chemically binding to unwanted impurities or reaction products. Calcium oxide, in the form of lime, was often used for this purpose, since it could react with the carbon dioxide and sulfur dioxide produced during roasting and smelting to keep them out of the working environment.

Of the seven metals known in antiquity, only gold occurred regularly in native form in the natural environment. The others – copper, lead, silver, tin, iron and mercury – occur primarily as minerals, though copper is occasionally found in its native state in commercially significant quantities. These minerals are primarily carbonates, sulfides, or oxides of the metal, mixed with other components such as silica and alumina. Roasting the carbonate and sulfide minerals in air converts them to oxides. The oxides, in turn, are smelted into the metal. Carbon monoxide was (and is) the reducing agent of choice for smelting. It is easily produced during the heating process, and as a gas comes into intimate contact with the ore.

In the Old World, humans learned to smelt metals in prehistoric times, more than 8000 years ago. The discovery and use of the "useful" metals — copper and bronze at first, then iron a few millennia later — had an enormous impact on human society. The impact was so pervasive that scholars traditionally divide ancient history into Stone Age, Bronze Age, and Iron Age.

In the Americas, pre-Inca civilizations of the central Andes in Peru had mastered the smelting of copper and silver at least six centuries before the first Europeans arrived in the 16th century, while never mastering the smelting of metals such as iron for use with weapon-craft.

Tin and Lead

In the Old World, the first metals smelted were tin and lead. The earliest known cast lead beads were found in the Çatal Höyük site in Anatolia (Turkey), and dated from about 6500 BC, but the metal may have been known earlier.

Since the discovery happened several millennia before the invention of writing, there is no written record about how it was made. However, tin and lead can be smelted by placing the ores in a wood fire, leaving the possibility that the discovery may have occurred by accident.

Although lead is a common metal, its discovery had relatively little impact in the ancient world. It is too soft to be used for structural elements or weapons, except for the fact that it is exceptionally dense, making it ideal for sling projectiles. However, being easy to cast and shape, it came to be extensively used in the classical world of Ancient Greece and Ancient Rome for piping and storage of water. It was also used as a mortar in stone buildings.

Tin was much less common than lead and is only marginally harder, and had even less impact by itself.

Copper and Bronze

After tin and lead, the next metal to be smelted appears to have been copper. How the discovery came about is a matter of much debate. Campfires are about 200 °C short of the temperature needed for that, so it has been conjectured that the first smelting of copper may have been achieved in pottery kilns. The development of copper smelting in the Andes, which is believed to have occurred independently of that in the Old World, may have occurred in the same way. The earliest current evidence of copper smelting, dating from between 5500 BC and 5000 BC, has been found in Pločnik and Belovode, Serbia. A mace head found in Can Hasan, Turkey and dated to 5000 BC, once thought to be the oldest evidence, now appears to be hammered native copper.

By combining copper with tin and/or arsenic in the right proportions one obtains bronze, an alloy which is significantly harder than copper. The first copper/arsenic bronzes date from 4200 BC from Asia Minor. The Inca bronze alloys were also of this type. Arsenic is often an impurity in copper ores, so the discovery could have been made by accident; but eventually arsenic-bearing minerals were intentionally added during smelting.

Copper–tin bronzes, harder and more durable, were developed around 3200 BC, also in Asia Minor.

The process through which the smiths learned to produce copper/tin bronzes is once again a mystery. The first such bronzes were probably a lucky accident from tin contamination of copper ores, but by 2000 BC, we know that tin was being mined on purpose for the production of bronze. This is amazing, given that tin is a semi-rare metal, and even a rich cassiterite ore only has 5% tin. Also, it takes special skills (or special instruments) to find it and to locate the richer lodes. But, whatever steps were taken to learn about tin, these were fully understood by 2000 BC.

The discovery of copper and bronze manufacture had a significant impact on the history of the Old World. Metals were hard enough to make weapons that were heavier, stronger, and more resistant to impact-related damage than their wood, bone, or stone equivalents. For several millennia, bronze was the material of choice for weapons such as swords, daggers, battle axes, and spear and arrow points, as well as protective gear such as shields, helmets, greaves (metal shin guards), and other body armor. Bronze also supplanted stone, wood, and organic materials in all sorts of tools and household utensils, such as chisels, saws, adzes, nails, blade shears, knives, sewing needles and pins, jugs, cooking pots and cauldrons, mirrors, horse harnesses, and much more. Tin and copper also contributed to the establishment of trade networks spanning large areas of Europe and Asia, and had a major effect on the distribution of wealth among individuals and nations.

Casting bronze ding-tripods, from the Chinese *Tiangong Kaiwu* encyclopedia of Song Yingxing, published in 1637.

Early Iron Smelting

Where and how iron smelting was discovered is widely debated, and remains uncertain due to the significant lack of production finds. Nevertheless, there is some consensus that iron technology originated in the Near East, perhaps in Eastern Anatolia.

In Ancient Egypt, somewhere between the Third Intermediate Period and 23rd Dynasty (ca. 1100–750 BC), there are indications of iron working. Significantly though, no evidence for the smelting of iron from ore has been attested to Egypt in any (pre-modern) period. There is a further possibility of iron smelting and working in West Africa by 1200 BC. In addition, very early instances of carbon steel were found to be in production around 2000 years before the present in northwest Tanzania, based on complex preheating principles. These discoveries are significant for the history of metallurgy.

Most early processes in Europe and Africa involved smelting iron ore in a bloomery, where the temperature is kept low enough so that the iron does not melt. This produces a spongy mass of iron called a bloom, which then has to be consolidated with a hammer. The earliest evidence to date for the bloomery smelting of iron is found at Tell Hammeh, Jordan (), and dates to 930 BC (C14 dating).

Later Iron Smelting

From the medieval period, the process of direct reduction in bloomeries began to be replaced by an indirect process. In this, a blast furnace was used to make pig iron, which then had to undergo a further process to make forgeable bar iron. Processes for the second stage include fining in a finery forge and, from the Industrial Revolution, puddling. However both processes are now obsolete, and wrought iron is now hardly made. Instead, mild steel is produced from a bessemer converter or by other means including smelting reduction processes such as the Corex Process.

Base Metals

Cowles Syndicate of Ohio in Stoke-upon-Trent England, late 1880s. British Aluminium used the process of Paul Héroult about this time.

The ores of base metals are often sulfides. In recent centuries, reverberatory furnaces have been used. These keep the fuel and the charge being smelted separate. Traditionally these were used for carrying out the first step: formation of two liquids, one an oxide slag containing most of the impurity elements, and the other a sulfide matte containing the valuable metal sulfide and some impurities. Such "reverb" furnaces are today about 40 m long, 3 m high and 10 m wide. Fuel is burned at one end and the heat melts the dry sulfide concentrates (usually after partial roasting), which are fed through the openings in the roof of the furnace. The slag floats on top of the heavier matte, and is removed and discarded or recycled. The sulfide matte is then sent to the converter. The precise details of the process will vary from one furnace to another depending on the mineralogy of the orebody from which the concentrate originates.

While reverberatory furnaces were very good at producing slags containing very little copper, they were relatively energy inefficient and produced a low concentration of sulfur dioxide in their off-gases that made it difficult to capture, and consequently, they have been supplanted by a new generation of copper smelting technologies. More recent furnaces have been designed based upon bath smelting, top jetting lance smelting, flash smelting and blast furnaces. Some examples of bath smelters include the Noranda furnace, the Isasmelt furnace, the Teniente reactor, the Vunyukov smelter and the SKS technology to name a few. Top jetting lance smelters include the Mitsubishi smelting reactor. Flash smelters account for over 50% of the world›s copper smelters. There are many more varieties of smelting processes, including the Kivset, Ausmelt, Tamano, EAF, and BF.

REFERENCES

1. Forrest C., 2008. The Nature of Scientific Experimentation in Archaeology: Experimental Archaeology from the Nineteenth to the Mid Twentieth Century. In P. Cunningham et al. (ed.) Experiencing Archaeology by Experiment 61-68. Oxford, Oxbow Books.
2. Craddock P. T., 1995. Early Metal Mining and Production. Edinburgh, Edinburgh University Press.
3. Mathieu J. R., 2002. Introduction – Experimental Archaeology: Replicating Past Objects, Behaviors, and Processes. In J. R. Mathieu (ed.) Experimental Archaeology: Replicating Past Objects, Behaviors, and Processes 1-11. Oxford, Archaeopress. BAR International Series 1023.
4. Pleiner R., 2000. Iron in Archaeology: The European Bloomery Smelters. Prague, Archaeology Ustav Avcr.
5. Killick D., 1991. The Relevance of Recent African Iron-Smelting Practice to Reconstructions of Prehistoric Smelting Technology. In P.D. Glumes (ed.) Recent Trends in Archaeometallurgical Research 47-54. Philadelphia, MASCA. MASCA Research Papers in Science and Anthropology Vol. 8 Part 1.
6. Craddock 1995, 22
7. Bernus S. and Echard N., 1985. Metal Working in the Agadez Region (Niger): An Ethno-Archaeological Approach. In P. T. Craddock and M. J. Hughes (ed.) Furnaces and Smelting Technology in Antiquity 71-80. London, British Museum. Occasional Papers No. 48.
8. Killick 1991
9. Timberlake S., 2007. The Use of Experimental Archaeology/Archaeometallurgy

for the Understanding and Reconstruction of Early Bronze Age Mining and Smelting Technologies. In S. L. Niece et al. (ed.) Metals and Mines: Studies in Archaeometallurgy 27-36. London, Archetype Press.

10. Crew P., 1990. Firesetting Experiment at Rhiw Goch. 1989. In P. Crew and S. Crew (ed.) Early Mining in the British Isles 57. Maentwrog, Plas Tan y Bwlch. Occasional Paper No. 1.

11. Lewis A., 1990. Firesetting Experiments on the Great Orme, 1989. In P. Crew and S. Crew (ed.) Early Mining in the British Isles 55-56. Maentwrog, Plas Tan y Bwlch. Occasional Paper No. 1.

12. Pickin J. and Timberlake S., 1988. Stone Hammers and Firesetting. Bulletin of the Peak District Mines Historical Society Ltd, Vol. 10 No. 3. 165-167.

13. Timberlake S., 1990. Firesetting and Primitive Mining Experiments, Cwmystwyth, 1989. In P. Crew and S. Crew (ed.) Early Mining in the British Isles 53-54. Maentwrog, Plas Tan y Bwlch. Occasional Paper No. 1.

14. Timberlake 2007

15. Crew 1990

16. Lewis 1990

17. Pickin and Timberlake 1988

18. Timberlake 1990

19. Timberlake 2007, 30

20. Craddock B., 1990. The Experimental Hafting of Stone Mining Hammers. In P. Crew and S. Crew (ed.) Early Mining in the British Isles 58. Maentwrog, Plas Tan y Bwlch. Occasional Paper No. 1

21. Coles J., 1979. Experimental Archaeology. London et. al, Academic Press

22. Bamberger M., 1985. The Working Conditions of the Ancient Copper Smelting Process. In P. T. Craddock and M. J. Hughes (ed.) Furnaces and Smelting Technology in Antiquity 151-157. London, British Museum. Occasional Papers No. 48.

23. Bamberger M. and Wincierzt P., 1990. Ancient Smelting of Oxide Copper Ore. In R. Rothenberg (ed.) The Ancient Metallurgy of Copper 78-122. London, Institute for Archaeo-Metallurgical Studies.

24. Merkel J. F., 1990. Experimental Reconstruction of Bronze Age Copper Smelting Based on Archaeological Evidence from Timna. In R. Rothenberg (ed.) The Ancient Metallurgy of Copper 78-122. London, Institute for Archaeo-Metallurgical Studies.

25. Tylecote R. F., 1991. Summary of Result of Experimental Work on Early Copper Smelting. . In W.A. Oddy (ed.) Aspects of Early Metallurgy 5-12. London, British Museum. Department of Science Occasional Paper No. 17.

26. Tylecote R. F., and Boydell P. J., 1978. Experiments on Copper Smelting Based on Early Furnaces Found at Timna. In B. Rothenberg (ed.) Chalcolithic Copper Smelting: Excavation and Experiments. London, Institute for Archaeo-

Metallurgical Studies. Archaeo-Metallurgy Monograph No. 1.

27. Tylecote R. F. and Merkel J. F., 1985. Experimental Smelting Techniques: Achievements and Future. In P. T. Craddock and M. J. Hughes (ed.) Furnaces and Smelting Technology in Antiquity 3-20. London, British Museum. Occasional Papers No. 48.
28. Zwicker U., 1991. Investigations on the Extractive Metallurgy of Cu/Sb/As Ore and Excavated Smelting Products From Norsun-Tepe (Keban) on The Upper Euphrates (3500-2800 B.C.). In W.A. Oddy (ed.) Aspects of Early Metallurgy 13-26. London, British Museum. Department of Science Occasional Paper No. 17.
29. Zwicker 1991
30. Craddock P., Meeks N., and S. Timberlake, 2007. On the Edge of Success: The Scientific Examination of the Products of the Early Mines Research Group Smelting Experiments. In S. L. Niece et al. (ed.) Metals and Mines: Studies in Archaeometallurgy 37-45. London, Archetype Press.
31. Tylecote 1991
32. Cushing F. H., 1894. Primitive Copper Working: an Experimental Study. American Anthropologist, Vol. 7, No. 1, 93-117.
33. Forrest 2008
34. Tylecote and Merkel 1985, 4
35. Cushing 1894
36. Pleiner 2000, 132
37. Tylecote and Merkel 1985, 8
38. Killick 1991, 47
39. Clough R. E., 1985. The Iron Industry in the Iron Age and Romano-British Period. In P. T. Craddock and M. J. Hughes (ed.) Furnaces and Smelting Technology in Antiquity 179-187. London, British Museum. Occasional Papers No. 48.
40. Coles 1979
41. Crew 1991
42. Crew P. and Charlton M. 2007. The Anatomy of a Furnace ... and Some of its Ramifications. In S. L. Niece et al. (ed.) Metals and Mines: Studies in Archaeometallurgy 219-225. London, Archetype Press.
43. ^ Jump Nosek 1985
44. Pleiner 2000
45. Clough 1985
46. 75. Tylecote and Merkel 1985, 13
47. Nosek E., 1985. The Polish Smelting Experiments in Furnaces with Slag Pits. In P. T. Craddock and M. J. Hughes (ed.) Furnaces and Smelting Technology in Antiquity 165-177. London, British Museum. Occasional Papers No. 48.
48. Tylecote and Boydell 1978
49. Bamberger and Wincierzt 1990

50. Merkel 1990

51. Cradock, Meeks, and Timberlake 2007

52. Crew P., 1991. The Experimental Production of Prehistoric Bar Iron. Journal of the Historical Metallurgy Society, Vol. 25 No. 1 21-36.

53. Pollard A. M., Thomas R. G., and P. A. Williams, 1989. Some Experiments Concerning the Smelting of Arsenical Copper. In P. Budd (ed.) Archaeological Sciences 169-174. Oxford, Oxbow Books. Monograph No. 9.

54. Tylecote and Merkel 1985, 10-11

55. Hetherington R., 1991. Investigations into Primitive Lead Smelting and Its Products. In W.A. Oddy (ed.) Aspects of Early Metallurgy 27-40. London, British Museum. Department of Science Occasional Paper No. 17.

56. Bayley J., Crossley D., and M. Ponting, 2008. Metals and Metalworking: A Research Framework for Archaeometallurgy. Wakefield, The Charlesworth Group. Historical Metallurgy Society Occasional Publication No. 6.

57. Roberts B. and Ottaway B., 2003. The Use and Significance of Socketed Axes During the Late Bronze Age. European Journal of Archaeology, Vol. 6, No. 2, 119-140.

58. Greenfield H. J., 2002. Distinguishing Metal (Steel and Low-tin Bronze) from Stone (Flint and Obsidian) Tool Cut Marks on Bone: An Experimental Approach. In J. R. Mathieu (ed.) Experimental Archaeology: Replicating Past Objects, Behaviors, and Processes 35-54. Oxford, Archaeopress. BAR International Series 1023.

59. Mathieu J. R. and Meyer D. A., 1997. Comparing Axe Heads of Stone, Bronze, and Steel: Studies in Experimental Archaeology. Journal of Field Archaeology, Vol. 24, No. 3, 333-351.

60. Mathieu J. R. and Meyer D. A., 2002. Reconceptualizing Experimental Archaeology: Assessing the Process of Experimentation. In J. R. Mathieu (ed.) Experimental Archaeology: Replicating Past Objects, Behaviors, and Processes 73-82. Oxford, Archaeopress. BAR International Series 1023.

61. Saraydar and Shimada 1973. Experimental Archaeology: A New Outlook. American Antiquity, Vol. 38, No. 3, 344-350.

62. "Металлургия. in The Great Soviet Encyclopedia. 1979.

63. Oxford English Dictionary, accessed 29 January 2011

64. History of Gold. Gold Digest. Retrieved 2007-02-04.

65. E. Photos, E. (2010). The Question of Meteoritic versus Smelted Nickel-Rich Iron: Archaeological Evidence and Experimental Results(PDF). World Archaeology. **20** *(3): 403. doi:10.1080/00438243.1989.9980081. JSTOR 124562.*

66. W. Keller (1963) The Bible as History. p. 156. ISBN 0-340-00312-X

67. H.I. Haiko, V.S. Biletskyi. First metals discovery and development the sacral component phenomenon. // Theoretical and Practical Solutions of Mineral Resources Mining // A Balkema Book, London, 2015, p. 227-233..

68. Radivojević, Miljana; Rehren, Thilo; Pernicka, Ernst; *Šljivar*, Dušan; Brauns, Michael; Borić, Dušan (2010). On the origins of extractive metallurgy: New evidence from Europe. Journal of Archaeological Science. **37** *(11): 2775. doi:10.1016/j.jas.2010.06.012.*

69. Neolithic Vinca was a metallurgical culture Stonepages from news sources November 2007

70. B. W. Anderson (1975) The Living World of the Old Testament, p. 154, ISBN 0-582-48598-3

71. R. F. Tylecote (1992) A History of Metallurgy ISBN 0-901462-88-8

72. Robert K.G. Temple (2007). The Genius of China: 3,000 Years of Science, Discovery, and Invention (3rd edition). London: André Deutsch. pp. 44–56. ISBN 978-0-233-00202-6.

73. Karl Alfred von Zittel (1901). History of Geology and Palaeontology. p. 15. doi:10.5962/bhl.title.33301.

74. Arthur Reardon (2011), Metallurgy for the Non-Metallurgist (2nd edition), ASM International, ISBN 978-1-61503-821-

75. Kuroda, Seiji; Kawakita, Jin; Watanabe, Makoto; Katanoda, Hiroshi (2008). Warm spraying—a novel coating process based on high-velocity impact of solid particles. Sci. Technol. Adv. Mater. **9** *(3): 033002. doi:10.1088/1468-6996/9/3/033002. PMC 5099653.*

76. Paulussen, S; Rego, R; Goossens, O; Vangeneugden, D; Rose, K (2005). Plasma polymerization of hybrid organic–inorganic monomers in an atmospheric pressure dielectric barrier discharge. Surface and Coatings Technology. **200**: 672. doi:10.1016/j.surfcoat.2005.02.134.

77. Leroux, F; Campagne, C; Perwuelz, A; Gengembre, L (2008). "Fluorocarbon nano-coating of polyester fabrics by atmospheric air plasma with aerosol". Applied Surface Science. **254** *(13): 3902. doi:10.1016/j.apsusc.2007.12.037.*

78. Moridi, A.; Hassani-Gangaraj, S. M.; Guagliano, M.; Dao, M. (2014). Cold spray coating: review of material systems and future perspectives. Surface Engineering. **30** *(6): 369–395. doi:10.1179/1743294414Y.0000000270.*

79. Degitz, Todd; Dobler, Klaus. Welding Journal. American Welding Society. Retrieved 5 August 2014.

80. Blunt, Jane and Balchin, N. C. (2001). Health and safety in welding and allied processes. Woodhead Publishing. pp. 190–205. ISBN 1-85573-538-5.

81. Suryanarayanan, R. (1993). Plasma Spraying: Theory and Applications. World Scientific Pub Co Inc. p. 211. ISBN 978-981-02-1363-3.

82. Bemer, D.; Regnier, R.; Subra, I.; Sutter, B.; Lecler, M. T.; Morele, Y. (2010). Ultrafine Particles Emitted by Flame and Electric Arc Guns for Thermal Spraying of Metals. Annals of Occupational Hygiene. **54** *(6): 607. doi:10.1093/*annhyg/meq052. PMID 20685717.

Index

U

V

W

X